Chen-Ning Yang

World Scientific Collection

Chen-Ning Yang

World Scientific Collection

Editor

KK Phua

World Scientific Publishing Company, Singapore

World Scientific

NEW JERSEY • LONDON • SINGAPORE • BEIJING • SHANGHAI • HONG KONG • TAIPEI • CHENNAI • TOKYO

Published by

World Scientific Publishing Co. Pte. Ltd.
5 Toh Tuck Link, Singapore 596224
USA office: 27 Warren Street, Suite 401-402, Hackensack, NJ 07601
UK office: 57 Shelton Street, Covent Garden, London WC2H 9HE

British Library Cataloguing-in-Publication Data
A catalogue record for this book is available from the British Library.

Compilation of CN Yang papers and related papers previously published by WSPC journals.

CHEN-NING YANG
World Scientific Collection

ISBN 978-981-98-2662-9 (hardcover)
ISBN 978-981-98-2663-6 (ebook for institutions)
ISBN 978-981-98-2664-3 (ebook for individuals)

For any available supplementary material, please visit
https://www.worldscientific.com/worldscibooks/10.1142/14686#t=suppl

Desk Editors Carmen Teo Bin Jie/Amalina Lai

Typeset by Stallion Press
Email: enquiries@stallionpress.com

World Scientific Titles by/about C. N. Yang

The Living Record of Scientific History: Conversations with CN Yang
ISBN: 978-981-12-8493-9

A Festschrift in Honor of the C N Yang Centenary: Scientific Papers
ISBN: 978-981-12-6414-6

Proceedings of the Conference in Honor of C N Yang's 85th Birthday: Physics, High Energy, Condensed Matter and Mathematical Physics
978-981-279-417-8

Selected Papers of Chen Ning Yang II: With Commentaries
ISBN: 978-981-4449-00-7
ISBN: 978-981-4449-01-4 (pbk)

Braid Group, Knot Theory and Statistical Mechanics
ISBN: 978-9971-5-0828-9
ISBN: 978-9971-5-0833-3 (pbk)

Braid Group, Knot Theory and Statistical Mechanics II
ISBN: 978-981-02-1524-8

Topology and Physics
ISBN: 978-981-327-849-3
ISBN: 978-981-327-850-9 (pbk)

Current Topics in Physics
(In 2 Volumes)
Proceedings of the Inauguration Conference of the Asia-Pacific Center for Theoretical Physics
ISBN: 978-981-02-3288-7

Differential Geometric Methods in Theoretical Physics
Proceedings of the XXI International Conference
ISBN: 978-981-02-1210-0

Frontiers of Science: In Celebration of the 80th Birthday of C N Yang
ISBN: 978-981-238-407-2
ISBN: 978-981-238-414-0 (pbk)

The Oskar Klein Memorial Lectures
Vol 1: Lectures by C N Yang and S Weinberg
ISBN: 978-981-02-0352-8
ISBN: 978-981-02-0353-5 (pbk)

60 Years of Yang–Mills Gauge Field Theories: C N Yang's Contributions to Physics
ISBN: 978-981-4725-54-5
ISBN: 978-981-4725-55-2 (pbk)

Lee Kuan Yew Through the Eyes of Chinese Scholars
ISBN: 978-981-320-231-3
ISBN: 978-981-320-937-4 (pbk)

World Scientific Titles by/about C. N. Yang (Chinese Publications)

杨振宁和翁帆的精选访谈
ISBN: 9789814139670 (pbk)

杨振宁《科学史论文选集》
ISBN: 9780962511875
ISBN: 9781879771048 (pbk)

杨振宁和曾帆谈美
ISBN: 9789814139908

六十八年新路
ISBN: 9789814632546

杨振宁的《黎明卷》
ISBN: 9789814139915
ISBN: 9789814139892 (pbk)

百年科学王氏: 阳真凝芳谈录
ISBN: 9789811277573 (pbk)

南大杨振宁教授
ISBN: 9789814139847

杨振宁访谈录及文章选集
ISBN: 9789623620017
ISBN: 9789623620024 (pbk)

杨振宁《黎明卷 (增订版)》
ISBN: 9789812839947 (pbk)

陈希吉
ISBN: 9789813278066
ISBN: 9789813278073 (pbk)

https://doi.org/10.1142/9789819826636_fmatter

Preface

In Memory of Professor Chen-Ning Yang

The passing of Professor Chen-Ning Yang is deeply mourned by all who had the privilege of learning from him or knowing him personally. A towering figure in twentieth-century physics, he was not only a brilliant scientist but also a lifelong champion of science, education, and scholarship in Asia.

In the 1970s, while I was a postdoctoral researcher at Stony Brook University, I had the great honor of studying under Professor Yang. His intellectual rigor and moral integrity left a profound and lasting impression on me, shaping my own journey in science and publishing.

In the early 1980s, as I observed how Western publishers dominated global academia, I began to dream of creating an academic press that would give Asia a stronger voice. Professor Yang wholeheartedly supported this vision. With his encouragement — and that of another Nobel Laureate, Professor Abdus Salam — my wife, Doreen Liu, and I founded World Scientific Publishing in 1981. From the very beginning, Professor Yang served as our academic advisor, steadfastly supporting our growth and sharing our belief in the importance of an Asian platform for scientific exchange.

Over the years, he remained closely connected with us, sending personal messages for our 30th and 40th anniversaries. To celebrate his 100th birthday, we co-published A Century of Science: Conversations with Chen-Ning Yang with East China Normal University Press — one of many meaningful collaborations we shared over more than four decades.

Professor Yang's ties with Singapore were deep and enduring. As early as the 1950s, he was invited to teach at Nanyang University. He first visited Singapore in 1967 and later served as an external examiner for Nanyang University's Physics Department, contributing significantly to the development of higher education in the country.

He was also a close friend of Singapore's former Deputy Prime Minister, Dr. Goh Keng Swee, who often sought his counsel on education and affectionately called him Zhuang Yuan (状元). In 2006, when Nanyang Technological University established the School of Physical and Mathematical Sciences, Professor Yang graciously lent his name to the C.N. Yang Scholars Programme, which continues to inspire generations of young scientists.

Professor Yang embodied the finest qualities of both Eastern and Western traditions. His relentless pursuit of truth and beauty in science, and his lifelong dedication to nurturing young minds, will continue to inspire generations to come.

KK Phua
November 2025

https://doi.org/10.1142/9789819826636_fmatter

Contents

Biographical sketch of C. N. Yang and his career

https://doi.org/10.1142/9789819826636_0001

Brief overview of C. N. Yang's 13 important contributions to physics*

Yu Shi
Department of Physics, Fudan University, Shanghai 200433, China
yushi@fudan.edu.cn

Published 30 December 2015

We give a brief overview of Professor Chen Ning Yang's 13 important contributions to physics, especially his contributions to gauge theory. The great impact of his relevant papers is analyzed. Commentary is made on Yang's distinctive style and trailblazing role in the history of physics.

1. Introduction

Professor Chen Ning Yang published two volumes of *Selected Papers with Commentaries* in 1983 and 2014 respectively.[1,2] In the preface of the first volume, Yang quoted the following two lines from Tu Fu (712–770):

A piece of literature
Is meant for the millennium
But its ups and downs are known
Already in the author's heart

On this volume of Yang, Freeman Dyson commented:[3]

Some of the chosen papers are important and others are unimportant. Some are technical and others are popular. Every one of them is a gem. Frank was not trying to cram as much hard science and possible into five hundred pages. He was trying to show us in five hundred pages the spirit of a great scientist, and he magnificantly succeeded.

A question arises then: among all these papers, which are most important? This question was answered by a selection of 13 contributions carved on a black cube,

*Invited talk given at the Conference on 60 Years of Yang–Mills Gauge Theories, IAS, NTU, Singapore, 25–28 May 2015.

Reproduced from International Journal of Modern Physics A, Vol. 30, No. 36, 1530069 (2015), DOI: 10.1142/S0217751X15300690

Fig. 1. The black cube engraved with C. N. Yang's 13 contributions in four areas of physics. Upper picture: front and right sides. Lower picture: back and left sides.

which was a gift for Yang's 90th birthday in 2012 (Fig. 1). On the top of the cube, the above quote of Tu Fu is engraved. On the four vertical surfaces are listed Yang's 13 important contributions to four areas of physics: statistical mechanics, condensed matter physics, particle physics and field theory.

Among the 13 contributions, those on statistical mechanics and condensed matter physics are,

- 1952 Phase Transition,
- 1957 Bosons,
- 1961 Flux Quantization,
- 1962 Off-diagonal Long-range Order,
- 1967 Yang–Baxter Equation,
- 1969 Finite Temperature,

the contributions to particle physics phenomenology are

- 1956 Parity Nonconservation,
- 1957 T, C and P,
- 1960 Neutrino Experiment,
- 1964 CP Nonconservation,

and the contributions to field theory are

- 1954 Gauge Theory,
- 1974 Integral Formalism,
- 1975 Fiber Bundle.

A commentary on these 13 contributions was recently made in considering their historic contexts in physics, their scientific beauty, and the taste and style of Yang.[4] Here we make some analyses on their impacts on the development of the relevant subjects. We make use of the citation database in *Web of Science*. Note that all the citation data are as of May 21, 2015.

2. Citation Analysis

Yang's 1952 papers on phase transitions include one on magnetization of the two-dimensional Ising model,[5] as well as two on the lattice gas model with T. D. Lee.[6,7] The total number of citations to these three papers are 698, 1181 and 1337, respectively. The number of citations to these three papers in total each year is shown in Fig. 2. One can observe that there was a jump in 1960s, and that it has almost kept

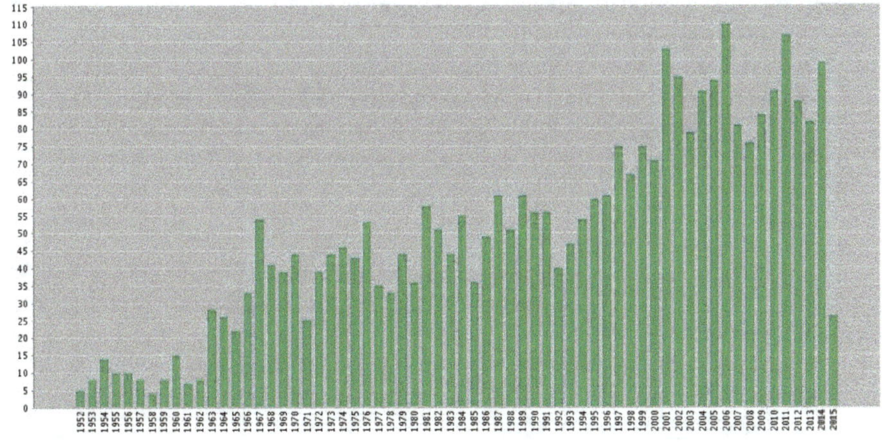

Fig. 2. The number of citations each year to Yang's three papers on phase transition of 1952. The sum is 3216.

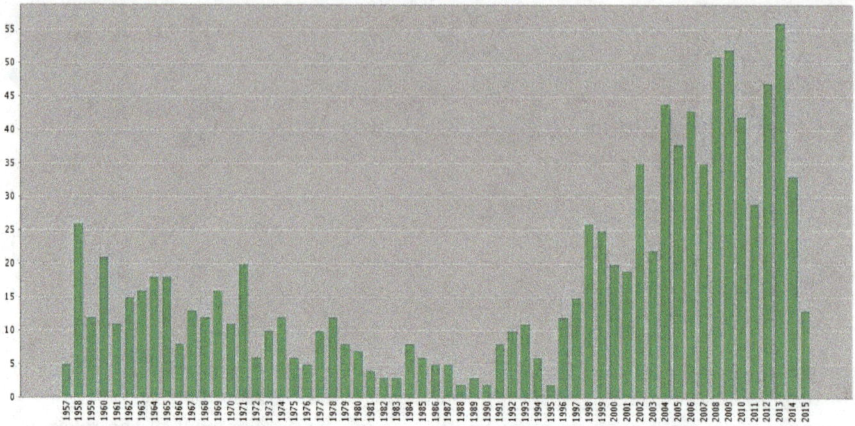

Fig. 3. The number of citations each year to Yang's two 1957 papers on dilute Bose gases. The sum is 1033.

increasing up to now. This is an "anomaly" in comparison with the usual trend that the number of citations each year eventually decreases with time.

Yang's 1957 papers on Bosons with his collaborators Kerson Huang and T. D. Lee[8,9] were among a series of papers on the dilute hard sphere Bose gas, which they chose as a mathematically well defined model yet motivated by the superfluidity of liquid helium. The total number of citations to these two papers is 643 and 390, respectively. The number of citations to these two papers in total each year is shown in Fig. 3. One can observe that it has kept increasing since the 1990s, apparently because of experimental breakthroughs brought about by modern techniques of cooling and trapping atoms.

Yang's 1961 paper with Byers on quantization of the magnetic flux trapped in a superconducting ring was a result of his interaction with solid state experimentalists at Stanford University.[10] The total number of citations to this paper is 702. The number of citations to this paper each year is shown in Fig. 4. One can see that there was a peak immediately following the publication, and there was a much more pronounced peak in 1990s and a plateau ever since, indicating great advance many years after the publication.

In 1962, Yang published his paper on off-diagonal long-range order as a unified framework of superfluidity and superconductivity.[11] The total number of citations to this paper is 932. The number of citations to this paper each year is shown in Fig. 5. It can be seen that there was a peak around 1968, moreover, there has been a plateau since the late 1980s, indicating much activity on quantum condensation phenomena in more recent years.

In 1967, after moving from Institute for Advanced Study at Princeton to Stony Brook, Yang published his paper on what has later been called Yang–Baxter

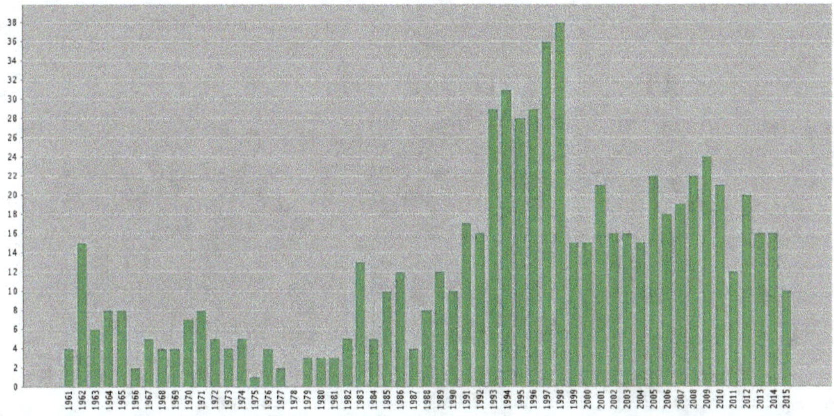

Fig. 4. The number of citations each year to Yang's 1961 paper on flux quantization of a superconducting ring. The sum is 702.

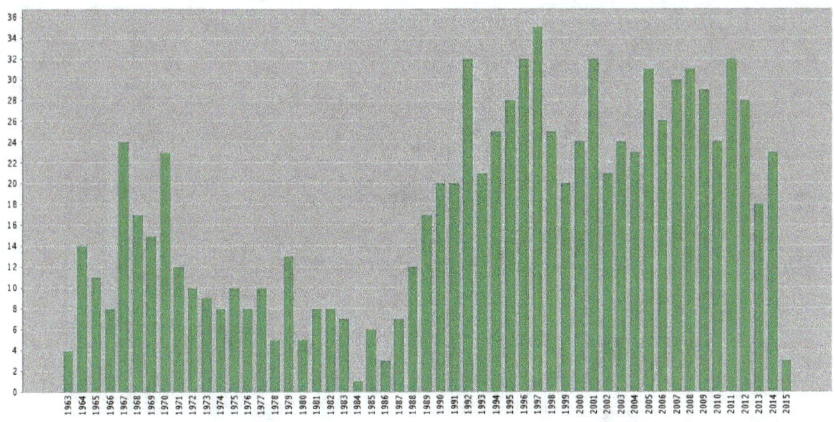

Fig. 5. The number of citations each year to Yang's 1964 paper on off-diagonal long-range order. The sum is 932.

equation.[12] As a theory of one-dimensional Fermions, it also provided a basis for analyzing many one-dimensional cold atom experiments of recent years. The total number of citations to this paper is 1393. The number of citations to this paper each year is shown in Fig. 6. It can be seen that there was a peak in 1980s and a much more pronounced peak in 1990s, and it has steadily increased since 2000.

In 1969, Yang published the exact solution of Bosons in one-dimensional repulsive potential at finite temperatures, with his brother C. P. Yang.[13] The model and its solution have also been realized and confirmed in cold atom experiments

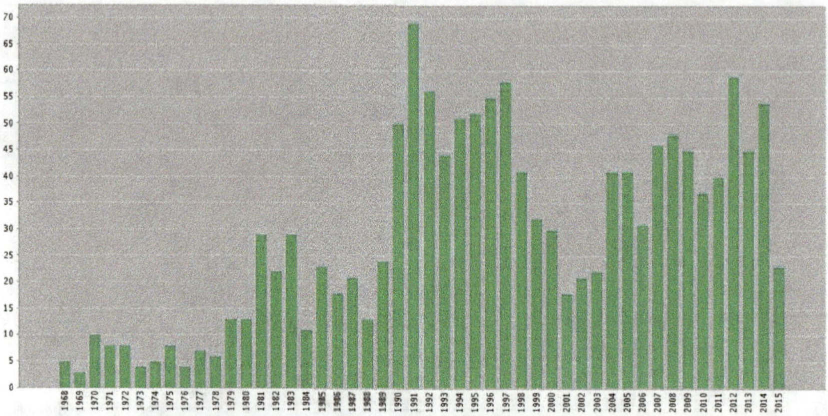

Fig. 6. The number of citations each year to Yang's 1967 paper on what has later been called Yang–Baxter equation. The sum is 1393.

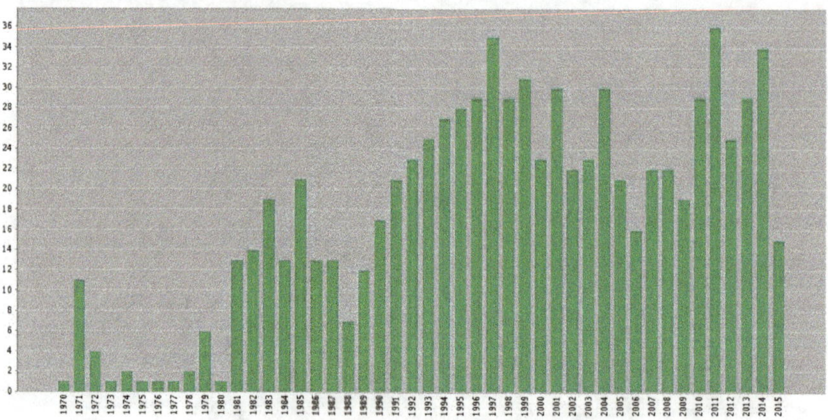

Fig. 7. The number of citations each year to the 1969 paper by Yang and Yang on one-dimensional Bosons at finite temperatures. The sum is 817.

in recent years. The total number of citations to this paper is 817. The number of citations to this paper each year is shown in Fig. 7. One can observe peaks in 1971, in 1980s, in 1990s, and in recent years. The peaks have become more and more pronounced.

Now we turn to Yang's papers on particle physics phenomenology. Yang's 1956 paper with T. D. Lee famously questioning whether parity is conserved in weak interactions earned them Nobel Prize in physics the very next year.[14] The speed of recognition is the highest one in all history of Nobel Prizes. The total number

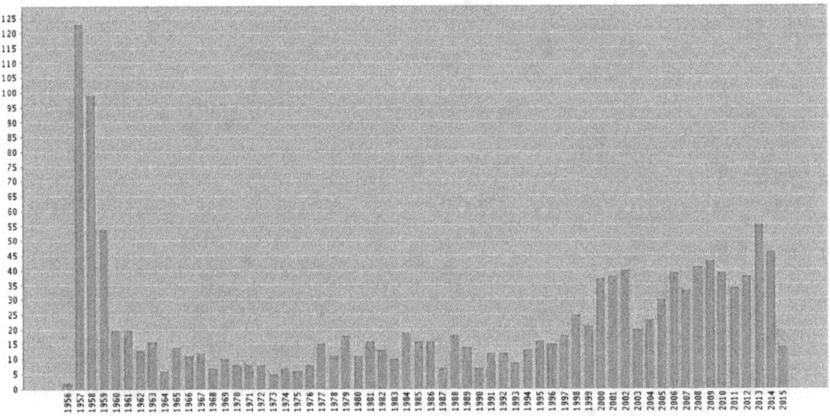

Fig. 8. The number of citations each year to the 1956 paper by Yang and Lee on parity violation in weak interactions. The sum is 1358.

of citations to this paper is 1358. The number of citations each year is shown in Fig. 8. The extremely high peak in 1957 was obviously because of the parity revolution. Interestingly, the citation each year has started increasing again since the 1990s.

In 1957, with Lee and R. Oehme, Yang published the paper on the relationship among violations of the discrete symmetries of time reversal (T), charge conjugation (C) and parity (P).[15] This paper had decisive impact on all theoretical analyses on CP violation later discovered in 1964. The total number of citations to this paper is 346. The number of citations to this paper each year is shown in Fig. 9. One can observe peaks following the parity revolution in 1957, and following the 1964 CP violation discovery, as well as in the 2000s.

In 1960, following the experimentalist M. Schwartz's idea that weak interactions at high energies can be studied by using neutrino beams, Lee and Yang explored theoretically the importance of such experiments,[16] which led to future development in neutrino physics. The total number of citations to this paper is 218. The number of citations to this paper in each year is shown in Fig. 10. There was a pronounced high peak immediately following the publication of this paper.

In 1964, Yang and T. T. Wu published their phenomenological analysis of the violation of CP invariance following its experimental discovery that year.[17] The number of citations to this paper is 278. The number of citations to this paper each year is shown in Fig. 11. There was a pronounced high peak immediately following the publication of this paper.

Now we come to Yang's greatest contribution to physics, that is, Yang–Mills gauge theory or non-Abelian gauge theory, which was founded in two short papers of 1954.[18,19] The first paper, which is less well known, was an abstract of Yang's

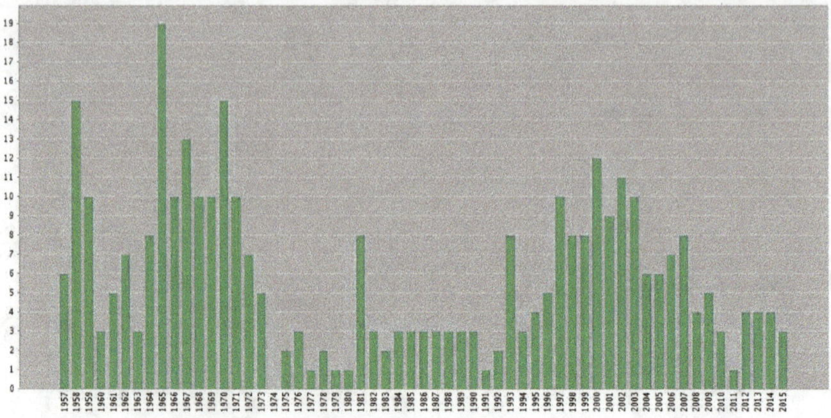

Fig. 9. The number of citations each year to the 1957 paper by Yang with Lee and Oehme on relationship among violations of T, C, and P in weak interactions. The sum is 346.

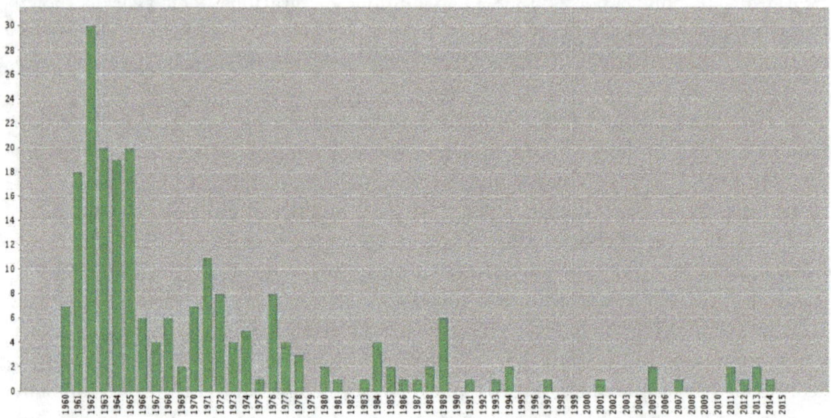

Fig. 10. The number of citations each year to the 1960 paper by Lee and Yang on high energy neutrinos. The sum is 218.

talk at the April meeting of the American Physical Society. The total numbers of citations to these two papers are 19 and 1740 respectively. The number of citations to these two papers in total each year is shown in Fig. 12. One can observe a huge peak in the 1970s, and it has kept increasing since 1990s. However, one should be cautioned that the original papers are often not cited in referring to Yang–Mills theory, just like that Einstein's 1905 paper is usually not cited in referring to special relativity. In *Web of Science*, for example, on May 21 2015, searching "Yang Mills"

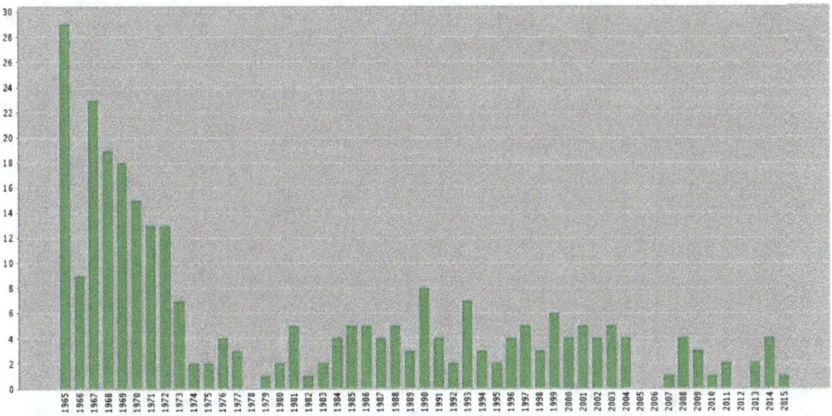

Fig. 11. The number of citations each year to the 1964 paper by Yang and Wu on CP violation. The sum is 278.

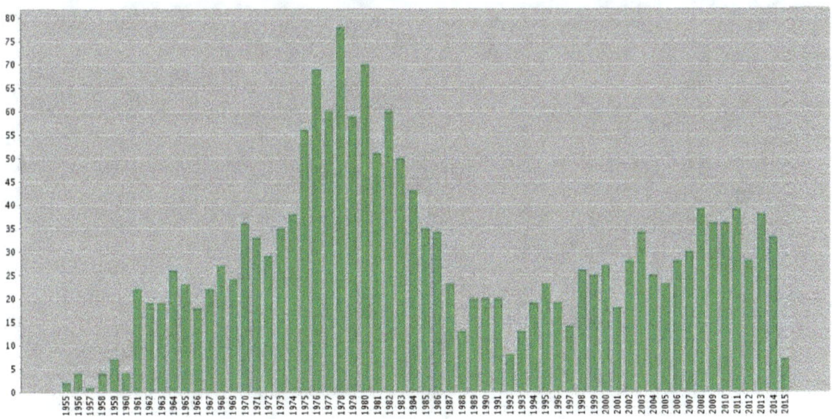

Fig. 12. The number of citations each year to the original 1954 papers by Yang and Mills on non-Abelian gauge theory. The sum is 1759.

in titles returned 5101 papers while searching "Yang Mills" as a topic returned 14967 papers.

In 1974, Yang published his paper on the integral formalism of gauge theory.[20] The total number of citations to this paper is 396. The number of citations to this paper each year is shown in Fig. 13. There was a pronounced peak following the publication of this paper.

In 1975, Yang and T. T. Wu published the paper on the correspondence between gauge theory and fiber bundle theory,[21] which inspired the subsequent reunion

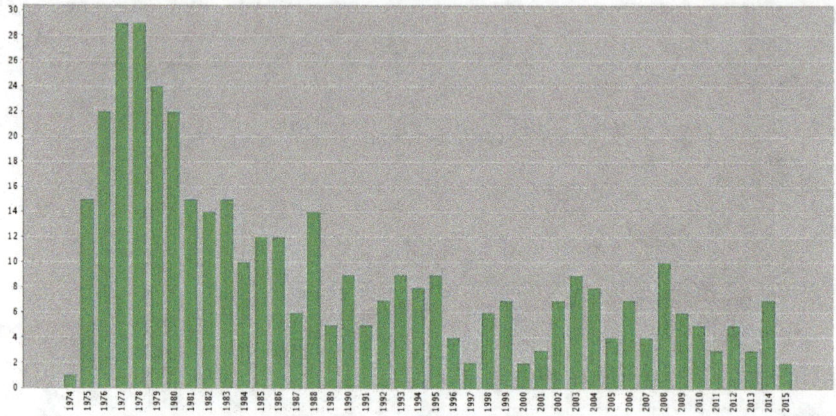

Fig. 13. The number of citations each year to the 1974 paper by Yang on integral formalism of gauge theory. The sum is 396.

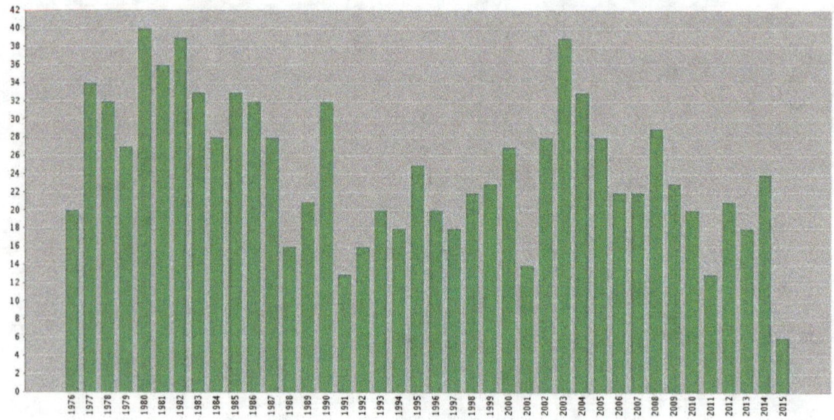

Fig. 14. The number of citations each year to the 1975 paper by Yang and Wu on the correspondence between gauge theory and fiber bundle theory. The sum is 993.

between mathematics and theoretical physics in last forty years. The number of citations to this paper is 993. The number of citations to this paper each year is shown in Fig. 14. There were two broad peaks in 1980s and in 2000s, without significant decay in between.

3. Two Types of Impact

We can classify the impact of these 13 contributions into two types.

Some of Yang's contributions were quickly confirmed by experiments or followed by others works, as indicated by immediate citation peaks. Moreover, citations may revive later because of new developments. Among the 13 important contributions, the following belong to this type:

- 1956 Parity nonconservation.
- 1957 T, C and P.
- 1960 Neutrino experiments.
- 1964 CP nonconservation.
- 1974 Integral formalism.
- 1975 Fiber bundle.

For the other nine important contributions, the importance may or may not have been recognized immediately following the publication, but they all received much more recognition after many years' further development. These contributions are:

- 1952 phase transition.
- 1957 Bosons.
- 1961 Flux quantization.
- 1962 ODLRO.
- 1967 Yang–Baxter equation.
- 1969 Finite temperature.
- 1954 Yang–Mills gauge theory.

Amongst these nine contributions, there are two extreme cases. For the 1952 phase transition, the number of citations each year has almost been increasing monotonically. For 1954 Yang–Mills gauge theory, many authors do not cite the original papers in referring to Yang–Mills theory.

Finally, we note that up to May 21, 2015, the total number of citations to the 16 papers representing the 13 contributions is 13867. The total number of citations to all the papers of Yang is 27124, with h-index 68. Moreover, at least 46 papers have been cited in 2015, including 15 among the 16 papers representing the 13 contributions. This indicates that the choice of the 13 contributions is indeed a good one.

The total number of citations each year to 13 contributions is shown in Fig. 15, while the total number of citations each year to all the papers by Yang is shown in Fig. 16. The two diagrams are similar in trend, indicating that the 13 contributions are indeed representative. In each of these two diagrams, there was a peak in 1957 because of parity revolution. Note that the number of citations each year is really the rate of the change of the total citations in all years up to that year. Thus its increase indicates the acceleration of the total number of citations up to that year. What is the "energy" driving this acceleration?

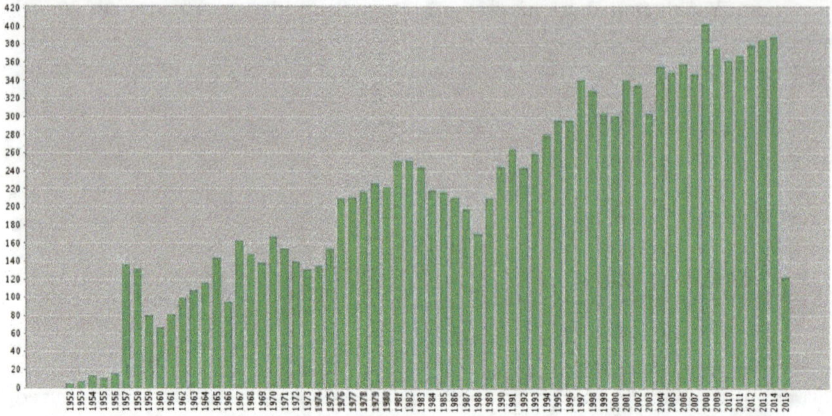

Fig. 15. The number of citations each year to the 16 papers representing the 13 contributions of by Yang. The sum is 13867.

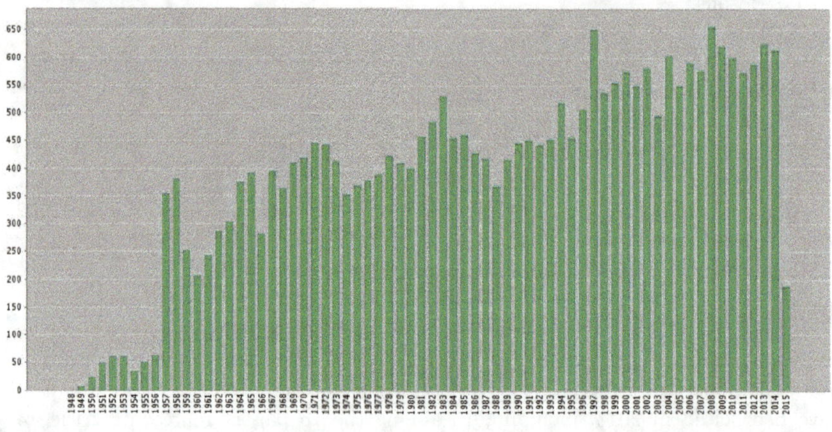

Fig. 16. The number of citations each year to all the papers of Yang. The sum is 27124.

4. Characteristic of Yang's Work

This leads us to the characteristics of Yang's work. Yang's work is distinguished by distinctive style and taste, which can be summarized as: *originality, elegance, power and physical relevance*. Yang has always paid great attention to experimental facts, basic structure of physics and beauty of theoretical form. He has always kept close to his own thoughts, ideas and intuition, rather than following fashions or jumping on bandwagons.

In the preface to his recently published collection of articles,[22] Dyson elaborates that his 1999 paper quoted above[3]

> *is a banquet speech, celebrating my friend Yang Chen Ning on the occasion of his retirement from Stony Brook in 1999. I called attention to three outstanding qualities of Yang that are rarely combined. First, a marvellous mathematical skill, enabling him to solve technical problems. Second, a deep understanding of nature, enabling him to ask important questions. Third, a community spirit, enabling him to play a major role in the rebirth of Chinese civilization. Together, these three qualities make him what he is, a conservative revolutionary who values the past and leads the way to the future.*

5. Yang–Mills Theory in Nobel Lectures

Yang–Mills theory, together with the ideas of spontaneous symmetry breaking and asymptotic freedom developed later by many authors, led to the Standard Model of particle physics, which has dominated research in fundamental physics in all subsequent years. The two 1954 papers by Yang and Mills opened the door of the principle "*Symmetry Dictates Interaction*", which has been the spirit of the main conceptual advance of theoretical physics in the last half century, and will continue to be general guidance for future development.

The building of the Standard Model has been recognized by many Nobel Prizes in physics. In the following, we make some quotations from the Nobel Lectures of these prizes.[23]

In his 1979 Nobel Lecture, Glashow said:

> *Today we have what has been called a "standard theory" of elementary particle physics in which strong, weak, and electromagnetic interactions all arise from a local symmetry principle.... Not only is electromagnetism mediated by photons, but it arises from the requirement of local gauge invariance. This concept was generalized in 1954 to apply to non-Abelian local symmetry groups.*

In his 1979 Nobel Lecture, Weinberg said:

> *The extension to more complicated groups was made by Yang and Mills in 1954 in a seminal paper in which they showed how to construct an SU(2) gauge theory of strong interactions.... To a remarkable degree, our present detailed theories of elementary particle interactions can be understood deductively, as consequences of symmetry principles and of a principle of renormalizability which is invoked to deal with the infinities.*

In his 1999 Nobel Lecture, Veltman recalled:

I therefore concluded that Yang–Mills theories are probably the best one can have with respect to renormalizability.... Starting then on the study of diagrams in a Yang–Mills theory I established the vanishing of many divergencies, provided the external legs of the diagrams were on the mass shell.

In his 1999 Nobel Lecture, 't Hooft recalled:

Back in 1971, I carried out my own calculations of the scaling properties of field theories, and the first theory I tried was Yang–Mills theory... Quantum chromodynamics, a Yang–Mills theory with gauge group SU(3), could therefore serve as a theory for the strong interactions.

't Hooft's prize-winning paper is entitled "Renormalizable Lagrangians for massive Yang–Mills fields".

In his 1999 Nobel Lecture, Gross recalled:

Gerard 't Hooft's spectacular work on the renormalizability of Yang–Mills theory reintroduced non-Abelian gauge theories to the community... We decided that we would calculate the β function for Yang–Mills theory... Politzer went ahead with his own calculation of the function for Yang–Mills theory... Our abstract reads: It is shown that a wide class of non-Abelian gauge theories have, up to calculable logarithmic corrections, free-field asymptotic behavior... His abstract reads: An explicit calculation shows perturbation theory to be arbitrarily good for the deep Euclidean Greens functions of any Yang–Mills theory and of many Yang–Mills theories with fermions...

In his 2008 Nobel Lecture, Nambu commented:

Thus the beautiful properties of electromagnetism was extended to the SU(2) non-Abelian gauge field.

In his 2008 Nobel Lecture, Kobayashi commented:

In the framework of the gauge theory, flavor mixing arises from mismatch between gauge symmetry and particle states.

In his 2008 Nobel Lecture, Maskawa recalled:

We started to study the CP violation based on the quartet quark model in the electroweak unified gauge theory.

In his 2013 Nobel Lecture, Englert said:

As a valid theory of short range interactions clearly required quantum consistency, we were naturally driven to take, as a model of the corresponding long range interactions, the generalization of quantum electrodynamics, known as Yang–Mills theory... To transmute long range interactions into short range ones in the context of Yang–Mills theory it would suffice to give these generalized photons a mass, a feature that, as we just indicated, is apparently forbidden by the local symmetries... The disappearance of the NG boson is thus an immediate consequence of local symmetry... Therefore the coupling of the would be NG boson to the gauge field must render the latter massive. This is the essence of the BEH mechanism.

In his 2013 Nobel Lecture, Higgs recalled:

Anderson remarked "The Goldstone zero-mass difficulty is not a serious one, because we can probably cancel it off against an equal Yang–Mills zero-mass problem."... Schwinger had, as recently as 1962, written papers in which he demolished the folklore that it is gauge invariance alone that requires photons to be massless. He had provided examples of some properties of a gauge theory containing massive "photons"... During the weekend 18–19 July it occurred to me that Schwinger's way of formulating gauge theories undermined the axioms which had been used to prove the Goldstone theorem. So gauge theories might save Nambu's programme.

6. Conclusion

In 2009, Dyson observed:[24]

Yang took Weyl's place as the leading bird among my generation of physicists... With non-Abelian gauge fields generating nontrivial Lie algebras, the possible forms of interaction between fields become unique, so that symmetry dictates interaction. This idea is Yang's greatest contribution to physics. It is a contribution of a bird flying high over the rain forest of little problems in which most of us spend our lives.

In 1999, Dyson assessed:[3]

The discovery of non-Abelian gauge fields was a laying of foundations for new intellectual structures that have taken 30 years to build. The nature of matter as described in modern theories and confirmed by modern experiments is a soup of non-Abelian gauge fields, held together by the mathematical symmetries that Yang first conjectured 45 years ago... Professor Yang is, after Einstein and Dirac, the preeminent stylist of the 20th century physics... He is a conservative revolutionary.

Let me conclude by adapting the two lines of Tu Fu:

Yang–Mills Theory
Is meant for the millennium
But its ups and downs are known
Already in the authors' heart

Acknowledgment

I thank Professor Chen Ning Yang for enlightening discussions.

References

1. C. N. Yang, *Selected Papers 1945–1980 with Commentary* (W. H. Freeman and Company, 1983); 2005 Edition (World Scientific, Singapore, 2005).
2. C. N. Yang, *Selected Papers II with Commentary* (World Scientific, Singapore, 2013).
3. F. Dyson, A Conservative revolutionary, *Int. J. Mod. Phys. A* **14**, 1455 (1999); also in A. Goldhaber, R. Shrock, J. Smith, G. Sterman, P. van Niuwenhuizen and W. Weisberger (eds.) *Symmetry and Modern Physics* (World Scientific, Singapore, 2003).
4. Y. Shi, *Int. J. Mod. Phys. A* **29**, 1475001 (2014). The author's talk at the Conference on 60 Years of Yang–Mills Gauge Field Theories was based partly on this article and partly on the present one.
5. C. N. Yang, The spontaneous magnetization of a two-dimensional Ising model, *Phys. Rev.* **85**, 808 (1952).
6. C. N. Yang and T. D. Lee, Statistical theory of equations of state and phase transitions. I. Theory of condensation, *Phys. Rev.* **87**, 404 (1952).
7. T. D. Lee and C. N. Yang, Statistical theory of equations of state and phase transitions. II. Lattice gas and Ising model, *Phys. Rev.* **87**, 410 (1952).
8. T. D. Lee and C. N. Yang, Many-body problem in quantum mechanics and quantum statistical mechanics, *Phys. Rev.* **105**, 1119 (1957).
9. T. D. Lee, K. Huang and C. N. Yang, Eigenvalues and eigenfunctions of a Bose system of hard spheres and its low-temperature properties, *Phys. Rev.* **106**, 1135 (1957).
10. N. Byers and C. N. Yang, Theoretical considerations concerning quantized magnetic flux in superconducting cylinders, *Phys. Rev. Lett.* **7**, 46 (1961).
11. C. N. Yang, Concept of off-diagonal long-range order and the quantum phases of liquid He and of superconductors, *Rev. Mod. Phys.* **34**, 694 (1962).
12. C. N. Yang, Some exact results for the many-body problem in one dimension with repulsive delta-function interaction, *Phys. Rev. Lett.* **19**, 1312 (1967).
13. C. N. Yang and C. P. Yang, Thermodynamics of a one-dimensional system of bosons with repulsive delta-function interaction, *J. Math. Phys.* **10**, 1115 (1969).
14. T. D. Lee and C. N. Yang, Question of parity conservation in weak interactions, *Phys. Rev.* **104**, 254 (1956).
15. T. D. Lee, R. Oehme and C. N. Yang, Remarks on possible noninvariance under time reversal and charge conjugation, *Phys. Rev.* **106**, 340 (1957).
16. T. D. Lee and C. N. Yang, Theoretical discussions on possible high-energy neutrino experiments, *Phys. Rev. Lett.* **4**, 307 (1960).
17. T. T. Wu and C. N. Yang, Phenomenological analysis of violation of CP invariance in decay of K^0 and $\bar{K}^0$, *Phys. Rev. Lett.* **13**, 380 (1964).

18. C. N. Yang and R. Mills, Isotopic spin conservation and a generalized gauge invariance, *Phys. Rev.* **95**, 631 (1954).
19. C. N. Yang and R. L. Mills, Conservation of isotopic spin and isotopic gauge invariance, *Phys. Rev.* **96**, 191 (1954).
20. C. N. Yang, Integral formalism for gauge fields, *Phys. Rev. Lett.* **33**, 445 (1974).
21. T. T. Wu and C. N. Yang, Concept of nonintegrable phase factors and global formulation of gauge fields, *Phys. Rev. D* **12**, 3845 (1975).
22. F. J. Dyson, *Birds and Frogs (Selected Papers, 1990–2014)* (World Scientific, Singapore, 2015).
23. http://nobelprize.org.
24. F. Dyson, Birds and frogs, *Notices of AMS* **56**(2), 212 (2009).

https://doi.org/10.1142/9789819826636_0002

Chen Ning Yang's New Contributions After He Returned to Where He Started*

Bang-Fen Zhu

Department of Physics and Institute for Advanced Study, Tsinghua University, Beijing 100084, China

Published 9 January 2018

Chen Ning Yang returned to Tsinghua University as a full professor in 2003. Regarding the fact that very few people know what Professor Yang has contributed to science and to China after his return, in this article new contributions of Chen Ning Yang are introduced as far as the author knows, including his leading role in China's sciences, the research in statistical physics, the role in cultivating gifted students, his research in history of science, and all other aspects relating to China's developments.

Chen Ning Yang was seven years old when he came to Tsinghua Campus. At that time, his father Wu-Chih Yang (Ko-Chuen Yang) worked as a professor in the Department of Mathematics, Tsinghua University. Department of Physics and Department of Mathematics in Tsinghua University at that time were in the Science Building, a place the young Chen Ning loved most, especially in summer, because it was particularly cool. In 2003, Prof. Yang returned to Tsinghua University as a full-time professor.[a] Later, the Centre for Advanced Study in Tsinghua University

*On the afternoon of August 21th, 2017, at the "Physics Frontier Conference" held by the Chern Institute of Mathematics, Nankai University, author presented a talk entitled "Chen Ning Yang's Contributions After He Returned to Where He Started." In accordance with the recording of the report, the Assistants to Editor of *Physics* produced a draft, and the author selected part of the content to be revised and finalized, which has been originally as Chinese published in *Physics* 2017 **47**(10). Part of the report that is not included in this article can be found in Refs. 1 and 2. This English version was translated by Ms. Li Yuying and Ms. Judy Yeo.

[a]In 1997, Chen Ning Yang promised Da-zhong Wang, the president of Tsinghua University then, to help Tsinghua develop basic science research. In 1999, Mr. Yang retired from Stony Brook University and accepted an invitation to be a full-time professor in Tsinghua University. His original plan was to settle down in Tsinghua University as soon as possible. However, because his wife, Chih-li Tu, was terminally ill and had to undergo treatment in the United States, he could not leave. In October 2003, after Tu passed away, he returned immediately.

Reproduced from International Journal of Modern Physics A, Vol. 33, No. 01, 1830001 (2018), DOI: 10.1142/S0217751X18300016

he founded moved back to the Science Building, where Prof. Yang described his life as that of a circle. During that time, he particularly enjoyed reading a poem by T. S. Eliot, a twentieth-century British poet, which he personally translated into Chinese. Two lines from the poem are: "In my beginning is my end /.../ In my end is my beginning" and "We shall not cease from exploration / And the end of all our exploring / Will be to arrive where we started / And know the place for the first time."

I was transferred to the Centre for Advanced Study in Tsinghua University in January 2000 as a full professor. Since then, I have also been a faculty member in the Department of Physics in Tsinghua University. I have the honor of having a lot of personal contacts with Prof. Yang. As far as I can see, Prof. Yang, who has come full circle, has become younger at heart, since his ending point becomes a new starting point. On his sixtieth birthday, Prof. Yang understood that "life is limited" and "this kind of idea had never appeared in my mind before I was 60 year-old" for the first time. In May 1999, at his emeritus banquet in the State University of New York at Stony Brook, Prof. Yang thought of Li Shangyin's poem "The sunset is endlessly endearing, (a pity) only that it is near the dusk," and Zhu Ziqing's sentence "As long as I can still witness the brilliant sunset, I don't want to mourn the dusk" to motivate himself.[3] However, in 2003 after he returned to Tsinghua, he wrote a poem *Root* with these two lines "耄耋新事业，东篱归根翁" [The octogenarian created a new career, when the Dong Li Weng returned hometown], expressing his desire to start a new career. In 2013, Prof. Yang published a new book *Selected Papers II With Commentaries*, in a commentary he adapted the words of Su Dongpo as following: "Who dares to say life is limited, Heaven has given me a second spring."[4] Obviously, his return in 2003 was a turning point. After his return, Prof. Yang began his new career and started the second spring in his life.

I once wrote an article "A few Chinese Masters of Physics I Have Known."[1] In this article, I described each master with one word. Regarding Prof. Yang, I wrote several words and pondered thoroughly, and decided to choose the word "frank." Prof. Yang's character is multi-faceted, why do I use the word "frank" to describe him? On the one hand, it is because of his frankness and sincerity, which may be found in his article "Father and Me." "I know, until the end of his life, he (Chen Ning Yang's father), in a corner of his heart, has never forgiven me for giving up my mother country." Chen Ning Yang and his father always had a strong bond, and Wu-Chih Yang never said a word about Chen Ning Yang's US citizenship, nor did he write anything. It was just Chen Ning Yang's own feelings. I think only a frank and honest person will reveal his internal monologue despite its private nature and the potential damage to his image. On the other hand, frankness indicates someone is innocent and outspoken. Through many years of personal contact with him, I really feel that Prof. Yang's psychological age is below his physiological age, farther below his real age, he truly has a heart that is "childlike and innocent."

Chen Ning Yang's comeback in 2003 was not like those who did not know the truth imagined, that it was for "pension" and "welfare." According to my observation, Prof. Yang, at the age of 80, started a new career and a new search, whereby made many new contributions simply to dedicate himself to the service of his motherland. In actual fact, from 80 to 95 years old, in those 15 years, the work he has done is not only far more than most scientific researchers, but also more important.

Prof. Yang's new contributions after the return can be summed up in five ways. I. The leadership role he played as a foresight scientist; II. The specific scientific research in the field of physics he has carried out as a physicist; III. The great contributions he has made in cultivating outstanding young talent for China as an educator; IV. A series of masterpieces he wrote as a science history researcher; V. Contributions in other fields. In view of the fact that many do not know the details, on the special occasion of Prof. Yang's 95 birthday, I would like to take this opportunity to introduce Prof. Yang's five new contributions, so that more people would more understand Chen Ning Yang.

1. Leader in China's Sciences

Helping to develop scientific research in China has been Prof. Chen Ning Yang's cherished wish for a long time. Prof. Yang was the first Chinese American scientist to visit China Mainland in July 1971 after more than 20 years of interruption in Sino-US relations. Since then, he has done a lot of practical work, trying his best to help China develop science and technology. After his return, there was more space and roles for him to play. In terms of organization and supervision of scientific research, he has accomplished four major things: The first was to successfully establish and lead Tsinghua University's Institute for Advanced Study; The second was to specify the direction for the development of Tsinghua University's Department of Physics, which fundamentally changed the aspects of Department of Physics; The third was to lay a foundation to Hong Kong Qiu Shi Science & Technologies Foundation and the Shaw Prize Foundation; The last was to provide insight and opinions on China's grand science projects.

As early as the time he was at graduate school in the National Southwest Associate University, Chen Ning Yang and his close friend Kun Huang seriously discussed the possibility of "successfully organizing a truly independent physical center" in China, and considered it to be more important than obtaining a Nobel Prize."[5] After Chen Ning Yang gained success and experience in the Institute for Advanced Study (IAS), Princeton, he had a dream: to create a theoretical research center in China similar to IAS, where several world-class theorists could lead a group of talented young scientists to do world-leading research in theoretical physics and mathematics. In the early and mid-1970s, when the "Cultural Revolution" took place and intellectuals entered labor reform, he visited the motherland many times, witnessing university professors classifying different electric resistors in factories. He deeply felt that "in those years, the Chinese government's one-sided egalitarianism had ruined China's science."[6] At the banquet on his second return to China, he outspokenly called on Premier Zhou Enlai to pay attention to China's fundamental sciences. Although this request was positively responded by Premier Zhou, reality was far from his dream. After the Cultural Revolution, the motherland ushered in a spring of science and began to attach importance to basic science research. In the 1980s, Prof. Yang actively helped to establish the Center for Advanced Study at Sun Yat-Sen University and the Theoretical Physics Division at the Chern Institute of Mathematics, Nankai University. Both institutions were successful, but in his mind both had not reached the level of IAS yet. In 1997, Chen Ning Yang accepted Tsinghua University's request to create the Center for Advanced Study, Tsinghua University (CASTU), which is now renamed as the Institute of Advanced Study, Tsinghua University, (IASTU), and to serve as the Honorary Director. After many discussions with Tsinghua University leaders, both sides decided to establish the CASTU based on the template model of Princeton's IAS.

As the leader of a scientific organization, Chen Ning Yang has given full attention to both the overall and strategic picture as well as details of important issues. As a man who does great things, Prof. Yang established Foundation of Tsinghua University Center for Advanced Study Co. Ltd., which was registered in Hong Kong and raised funds in Hong Kong and the United States shortly after the CASTU was founded. Prof. Yang understood that with financial freedom, it was possible to have the freedom to recruit brilliant scientists and to open new research directions. Therefore, on the one hand, Prof. Yang donated his personal savings and a 3-acre villa on Long Island in the United States to the Foundation. On the other hand, he also tried his best to raise funds among friends and patriots in Hong Kong. It is through this Foundation that it became possible for the Institute of Advanced Study to induct Prof. Andrew Chi-Chih Yao, a Member of the National Academy of Sciences and winner of the A. M. Turing Award, as a full-time professor in Tsinghua.

Only with the Foundation was it possible to set up the "Chen Ning Yang Professor" position, which provided Weng Zhengyu, Wang Xiaoyun, and a batch of outstanding young scientists with decent salaries. And it also meant better treatment for doctoral and post-doctoral researchers. Not only did Prof. Yang donate the founding sum of the Foundation, he also donated a large part of his subsidy which was given to him after his return, of one million yuan per year to help improve the operations of the IASTU.

A correct choice of research directions and the recruitment of first-class researchers are keys to making a scientific institute successful. The research direction that Prof. Yang initially set for the Center lay in theoretical physics and mathematics. Specifically, the focus of theoretical physics at the Center was on condensed matter physics and cold atom physics. For this purpose, a number of the best Chinese scholars in international physics community were introduced to the Center including Shou-Cheng Zhang, Xiao-Gang Wen, Dung-Hai Lee, Jason Tin-Lun Ho, Terrence Hwa. They made important contributions to the Center. In 2005, inspired by Prof. Yang, Prof. Yao, a master in the field of computer theory, joined CASTU as a full-time professor. Subsequently, Wang Xiaoyun, an outstanding female scientist in the field of mathematics and cryptography, joined CASTU as well. Prof. Yang also invited distinguished scholars at Microsoft Research Asia as Adjunct Professors, and computer science was included in the fields of research at the Center. Prof. Yang put a lot of effort into recruiting talented people to work at the Center. For each candidate, Prof. Yang had to carefully study his or her academic background and achievements. Often, he talked to many candidates, each for several rounds, and finally selected only one. In terms of the composition of researchers and administrative operational mechanisms, the CASTU has taken a similar approach to that of the Institute of Advanced Study at Princeton. In line with the principle of "highly capable, excellent, and mobile," it actively recruits scientists at home and abroad through various channels by creating a relaxed environment, encouraging immersive scientific exploration, cultivating top talent. In the past 20 years, the IASTU has become an "academic palace." The world has witnessed continuous reports of first-class scholars gathered here, including Nobel Prize winners and a large number of young scholars who are active on the frontiers of physics. IASTU members work hard and learn from each other, leading to multiple cases of cross-disciplinary integration, tremendous academic achievements and increasing numbers of academic elites. It had profound influence on the development of Tsinghua's scientific disciplines. Furthermore, as Jining Chen, former president of Tsinghua University, pointed out: "The IASTU, which was established by Chen Ning Yang, not only made important contributions to frontier academic research but also exerted far-reaching influence on Tsinghua University's reform in administration and systems, which has played an important role in transforming Tsinghua into a world-class university."

Prof. Yang played a key role in remodelling the Department of Physics, Tsinghua University. In June 2002, President Da-zhong Wang invited Chen Ning Yang, Yuen-Ron Shen, Zhi-Xun Shen and Ping Sheng to form an International Advisory

Committee for Department of Physics, and launched the first international evaluation for a department in the history of Tsinghua University. The four Committee members, "Three Shens and One Yang," spent two days listening to comprehensive reports of each research group, visiting a number of laboratories, exchanging in-depth views with the University and Department administrators, professors, academicians, discussing with young teachers, students, laboratory technicians, office staffs, and members of Department Affairs Committee and Teaching Committee. Through careful considerations and many private communications, the Committee submitted an evaluation report to University. The international assessment report caused a lot of controversy and chatter in the Tsinghua campus. University administrators asked the Department of Physics to set up a special working group for the implementation of proposals in the assessment report. I was a professor at the CASTU then. It was exactly during the discussions on the evaluation report that I started to be involved in the affairs of the Department of Physics. After I became Head of the Department of Physics in April of 2003, I demanded that the evaluation report be released to all faculties of the Department. This caused much discussion and even disputes in the Department. Many had never seen an evaluation report that pointed out existing problems so sharply and so clearly indicated the direction of development. Some teachers thought the report was too sensitive to be made openly available. However, I thought this report was crucial to the future of Tsinghua Physics, and I finally made it available to the entire Department. The most important feature of the report is that it gives clear guidelines for the development of Tsinghua Physics. In particular, as expressed in the report: Firstly, "experimental research needs to be strengthened." Secondly, "we agree upon choosing condensed matter physics as the predominant discipline. But current experimental research in this field is rather weak. Nanomaterials growth and superconductor applications mostly belong to chemistry, materials and engineering, with research in physical properties constituting a very small part. The Department of Physics should enhance investigating physical properties by including different fields of condensed matter physics and promoting cooperation and exchanges between theory and experiment." "There are so many difficulties in the future development of high-energy physics and nuclear physics that it is not appropriate to hire more faculty members in these two fields." Thirdly, "The Department should create an environment which priorities teaching." "It should be clarified once more that teaching is still a university's most important duty, even in current times when the scientific research is emphasized unprecedentedly." "Every professor, under normal circumstances, must teach in each semester, which can help promote the unity between teaching and research. Professors with outstanding scientific research achievements should be encouraged to teach basic courses. That teaching is an important part of assessment should be clearly stipulated." These three points made outline for the development of Tsinghua's Department of Physics and played key roles in adjusting plans and the strategic approach for the academic discipline as well as key developments.

Fig. 1. A photo of Prof. Yang (middle), Qi-Kun Xue (left) and Yayu Wang (right).

In March of 2013, Qi-Kun Xue's research group in this Department observed the quantum anomalous Hall effect experimentally for the first time in the world. I think there would be no such research breakthrough without Prof. Yang. For on the one hand, Prof. Yang invited Shou-Cheng Zhang to visit the CASTU regularly, and Shou-Cheng introduced the emerging field of topological insulators to two excellent Tsinghua PhD candidates then, Xiao-liang Qi and Chao-xing Liu. And Xiao-liang, Chao-xing and their collaborators have theoretically predicted the experiment mechanism and practical material system for observing the quantum anomalous Hall effect. On the other hand, Qi-Kun Xue joined Tsinghua Physics in 2005, simply resulted from the first international evaluation for Tsinghua Physics led by Prof. Yang, which emphasized Tsinghua's Department of Physics should focus on the development of experimental condensed matter physics. It was Prof. Yang who provided a platform for cooperation between Shou-Cheng Zhang and Qi-Kun Xue that led to the experimental result of Xue's group for observing the quantum anomalous Hall effect. On hearing the good news, Prof. Yang was very happy and invited Qi-Kun Xue, Yayu Wang, and their wife to dinner immediately. Yayu Wang is one of Qi-Kun Xue's main collaborators. Prof. Yang gave an evaluation of their finding as "an achievement of roughly equal level to the Nobel Prize." Figure 1 is a photo of Prof. Yang, Qi-Kun Xue and Yayu Wang after the dinner. This example reflects Prof. Yang's foresight as a physics guru and leader.

From May 30th to June 1st of 2010, Tsinghua University conducted a second international assessment for the Department of Physics. The new committee consists of six members; "Four Shens and Two Yangs," i.e. in addition to the previous

"Three Shens and One Yang," there is new member: Lu Jeu Sham and Bing-Lin Yang. While pointing out the remaining problems of Tsinghua's Department of Physics as sharp as before, the assessment report concluded as: "Since the first assessment had been done in 2002, Tsinghua's Department of Physics has greatly improved in all aspects. Under the general principles of teaching and research ruled by the professors, scientific research strength has attained international standards. In some areas, it is even in the ranks of world leaders. Teaching is very successful too. Each year the Department of Physics can attract top-notch undergraduates in China who are quite satisfied with the teaching in this Department, a huge difference from eight years ago. If the Department of Physics makes further improvements on such basis, it has the chance to attain world-class standards. To achieve this goal would still require concerted effort from colleagues within the Department." Borrowing the words of former president Dazhong Wang: "Without Professor Yang, Tsinghua University's Department of Physics would not have these achievements today."

Prof. Yang is the Founding Council Member of the Hong Kong Shaw Prize Foundation and the founding consultant of Hong Kong Qiu Shi Science & Technologies Foundation. As a Founder he established the aims and high standards for these two prizes. Mr. Cha Chi-ming in Hong Kong wanted to help China to develop science and technology. In 1993 he told Chen Ning Yang, that the funds of the foundation could be provided by the Cha family, while the organization, operation and award principle and evaluation process of the Foundation would be taken on by Prof. Yang, an accomplished scientist who is concerned about China. Both decided that the purpose of the Qiu Shi Science & Technology Prize was in "providing timely help, rather than being blessed with a double portion of good fortune." Chen Ning Yang recalled the impoverished circumstances of Deng Jiaxian, one of his best friends, and reflected on the fact that China's most outstanding scientists were still working in extremely poor conditions. So in 1994, the Foundation decided to award 10 outstanding scientists in Mainland (including part of "The Hero of Atomic-Bomb, Hydrogen-Bomb and Satellite" like Deng Jiaxian, Yu Min and Zhou Guangzhao). In 1995, in order to encourage outstanding young scientists to remain in the motherland to do basic research, the Foundation set up the Qiu Shi Distinguished Young Scholars Award, and many award winners later became leaders in Chinese scientific research. In 1996, the Foundation granted Tu Youyou and other 10 researchers the "Qiu Shi Outstanding Scientific and Technological Team Award" in recognition of their achievements in researching artemisinin and its derivatives. At that time no organization granted Tu Youyou any award.

The Shaw Prize, which was set up in 2004, aims to confer awards in the fields of mathematics, astronomy, life sciences and medicine. As the first chairman of the Board of Adjudicators, Prof. Yang established the Shaw Prize is designed as a complementary prize of the highest level to the Nobel Prize. And because of Prof. Yang's hard working and his insight, the Shaw Prize has rapidly built up a reputation. Many world's most distinguished mathematicians, astronomers, doctors

Fig. 2. On September 14, 2017, Prof. Yang visited Shanghai Soft X-ray Free Electron Laser Device Facility (provided by Shanghai Synchrotron Radiation Facility).

and life scientists won the Shaw Prize, and respectable winners have won the Nobel Prize after obtaining the Shaw Prize.

Prof. Yang is very concerned about China's grand scientific projects as well as science and technology policies, and often shares his own opinions on these issues. He never sought any selfish interest for himself or for interest groups, and he focused solely on the development of science and technology in China. As early as in 1997, Prof. Yang was quite forward-looking in his suggestion that China should develop X-ray Free Electron Lasers which was raised on many occasions. In January 2017, the world's first fully-coherent Free Electron Laser in the wavelength range of 20–100 nm was established at the Dalian Institute of Chemical Physics, Chinese Academy of Sciences. A soft X-ray free electron laser was also commissioned at the Shanghai Institute of Applied Physics, which is also developing a larger hard X-ray Free Electron Lasers with high-repetition rate driven by superconducting accelerator at the same time. China's Free Electron Laser made a big leap forward through Prof. Yang's promotion of the cause (Fig. 2). Prof. Yang's opinion on the recent controversy over the construction of the super-grand collider in China clearly shows his love of China and his sincere wish to serve the Chinese people, which is well admitted by nearly every scientist regardless of the position he/she holds on that matter. With his academic achievements and reputation, Prof. Yang has played a leading role in the science community of China.

2. A Senior But Still Ambitious Physicist

On his 90th birthday, Tsinghua University sent him a black cube (Fig. 3) as a gift, engraved with lines from Prof. Yang's favorite poet Du Fu — "A piece of literature/Is meant for the millennium/But its ups and downs are known/Already in the

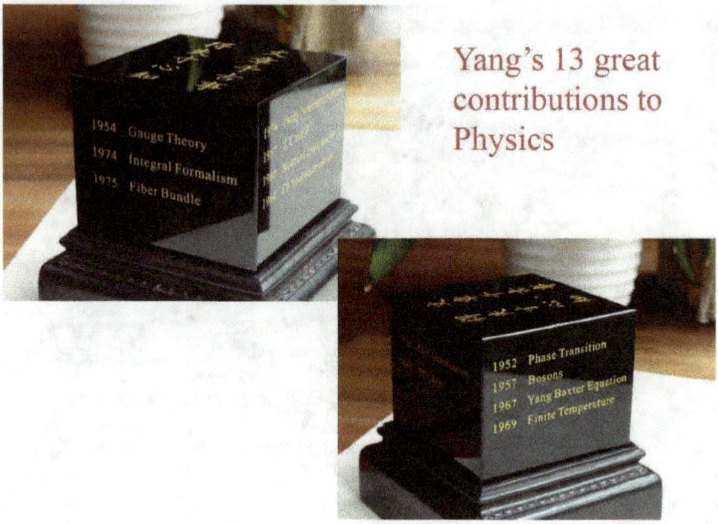

Fig. 3. The gift from Tsinghua University for Prof. Yang on his 90th birthday.

author's heart." And listed on the four sides were his 13 major contributions to physics in the four areas: field theory, particle physics, statistical physics and condensed matter physics. These contributions were made before he returned to China. When he settled down in Tsinghua Campus, Prof. Yang was already over eighty years old, but he still endeavored to work hard at the frontlines of research. Figure 4 shows the list of 27 SCI articles he published after he returned to Tsinghua. These articles can be divided into two categories, one comprises purely physics research articles, the other one contains some research articles related to physics history and the interpretation of some physics concepts. In addition, he has published two books and a large number of Chinese academic papers.

As for the selection of physical research topics, Prof. Yang proposed two principles based on his own experience: "finding topics that have a direct and simple connections with physical phenomenon, or with the basic structure of physics" and "expanding the problem often leads to a good new direction." These two principles served as good guidance for postgraduates and researchers. Chen Ning Yang still followed this "direct and simple connections" principle to determine his own research topics at Tsinghua. After returning, he mainly did research in the field of statistical physics, because, on the one hand statistical physics was always his favorite subject area, and helped realize the combination of the beauty of mathematics and the charm of physics which he admired; on the one hand, after entering this century, with the progress in laser cooling technology, cold atom physics has become one of the most active and important frontiers of physics research, and Prof. Yang's earlier

CN Yang's published 27 SCI papers at Tsinghua

1. Yang, C. N., Jeremy Bernstein's monologue, *Modern Physics Letters A* Vol. 32, Issue 19, 1730017 (2017)
2. Yang, C. N., China should not build a supercollider at this time, *International Journal of Modern Physics A* Vol. 31, Issue 30, 1630053 (2016)
3. Yang, C. N., The Future of Physics – Revisited, *International Journal of Modern Physics A* Vol. 30, Issue 21, 1530049 (2015)
4. Yang, C. N., C. S. Wu's Contributions: A Retrospective in 2015, *International Journal of Modern Physics A* Vol. 30, Issue 20, 1530050 (2015)
5. Yang, Chen Ning, The conceptual origins of Maxwell's equations and gauge theory, *Physics Today* Vol. 67, 45-51 (2014)
6. Wei, B. B.; Yang, C. N., Solvable lattice gas models with three phases, *EPL* Vol. 102, 66006 (2013)
7. Yang, Chen Ning, Topology and gauge theory in physics, *International Journal of Modern Physics A* Vol. 27, 1230035 (2012)
8. Yang, Chen-Ning, Weighing in on Wu's legacy, *Physics World* Vol. 25, 22 (2012)
9. Yang, C. N., My experience as student and researcher, *International Journal of Modern Physics A* Vol. 27, 1230009 (2012)
10. Yang, Chen Ning, Fermi's beta-decay theory, *International Journal of Modern Physics A* Vol. 27, Issues 3-4, 1230005 (2012)
11. Yang, Chen Ning, Quantum numbers, Chern classes, and a bodhisattva, *Physics Today* Vol. 65, Issue 1, 33-34 (Jan 2012)
12. Yang, C. N.; You Yi-Zhuang, One-Dimensional w-Component Fermions and Bosons with Repulsive Delta Function Interaction, *Chinese Physics Letters* Vol. 28, 020503 (2011)
13. Ma Zhong-Qi; Yang, C. N., Bosons or Fermions in 1D Power Potential Trap with Repulsive Delta Function Interaction, *Chinese Physics Letters* Vol. 27, 090505 (2010)
14. Ma Zhong-Qi; Yang, C. N., Spin 1/2 Fermions in 1D Harmonic Trap with Repulsive Delta function Interparticle Interaction, *Chinese Physics Letters* Vol. 27, 080501 (2010)
15. Yang, C N., Some problems in cold atom research, *International Journal of Modern Physics B* Vol. 24, 3469-3477 (2010)
16. Ma Zhong-Qi; Yang, C. N., Spinless Bosons in a 1D Harmonic Trap with Repulsive Delta Function Interparticle Interaction II: Numerical Solutions, *Chinese Physics Letters* Vol. 27, 020506 (2010)
17. Yang, C. N., Ground State of Fermions in a 1D Trap with Delta Function Interaction, *Chinese Physics Letters* Vol. 26, 120504 (2009)
18. Ma Zhong-Qi; Yang, C N., Spinless Bosons in a 1D Harmonic Trap with Repulsive Delta Function Interparticle Interaction I: General Theory, *Chinese Physics Letters* Vol. 26, 120506 (2009)
19. Ma ZQ; Yang, C N., Ground State Energy for Fermions in a 1D Harmonic Trap with Delta Function Interaction, *Chinese Physics Letters* Vol. 26, 120505 (2009)
20. Wei, Bo-Bo; Yang, Chen-Ning, Pair distribution function of one-dimensional "hard sphere" Fermi and Bose systems, *EPL* Vol. 87, 10005 (2009)
21. Weng, Z. Y.; Yang Chen Ning, A kinematic model for the universal T_c vs. doping curve in cuprates, *EPL* Vol. 85, 17005 (2009)
22. Yang. C.N., Pseudopotential method in cold atom research, in *Proc. of the 9th International Symposium on Foundations of Quantum Mechanics in the Light of New Technology*, Ed. by Ishioka, S; Fujikawa, K (World Scientific, 2009), pp. 3-4.
23. Yang, C. N., Pseudopotential method and dilute hard "sphere" Bose gas in dimensions 2, 4 and 5, *EPL* Vol. 84, 40001 (2008)
24. Wu, A. C. T.; Yang, Chen Ning, Evolution of the concept of the vector potential in the description of fundamental interactions, *International Journal of Modern Physics A* Vol. 21, 3235-3277 (2006)
25. Yang, Chen Ning; Ge, Mo-Lin, Bethe's hypothesis, *International Journal of Modern Physics B* Vol. 20, 2223-2225 (2006)
26. Yang, Chen Ning, Albert Einstein: Opportunity and perception, *International Journal of Modern Physics A* Vol. 21, 3031-3038 (2006)
27. Yang, CN, Thematic melodies of twentieth century theoretical physics: Quantization, symmetry and phase factor, *International Journal of Modern Physics A* Vol. 18, 3263-3272 (2003)

Fig. 4. List of 27 SCI articles published by Prof. Yang after he returned to Tsinghua University.

theoretical predictions in statistical physics have been confirmed through experiments, which would naturally lend an impetus to further theoretical research. The statistical physics research carried out after Yang's return is based on two idealized models whose physical structures are simple and direct: the dilute hard sphere bosons and the multi-particle boson system with one-dimensional repulsive potential (these two belong to his 13 major physics contributions). In the mid-1950s, Chen Ning Yang and his collaborators completed a series of studies on the dilute hard sphere boson many-body system due to their common interests in the superfluidity of liquid helium. They obtained the identical ground state energy with the double collision method and the pseudopotential method. Their most surprising result is that the ground state energy correction is proportional to the square root of the density, that is

$$\left(\frac{2m}{\hbar^2}\right)\frac{E_0}{N} = 4\pi a\rho\left[1+\frac{128}{15\sqrt{\pi}}\sqrt{\rho a^3}\right],$$

which at that time could not be verified experimentally. Fifty years later, this correction term has been experimentally confirmed with the development of cold atom physics. Prof. Yang re-studied this problem by himself, using the pseudopotential method to extend the dilute hard sphere boson many-body system from 3-dimensional to 2-dimensional, 4-dimensional and 5-dimensional.

In 1969, Chen Ning Yang and his brother C. P. Yang studied the boson problem in one-dimensional repulsive potential at finite temperature. This is the first rigorous solution of the quantum statistical model with interaction at finite temperature. This model is also built up and confirmed by the cold atom system. After the return, Prof. Yang extended it to the one-dimensional fermion subsystem, to the particle system with multiple components, and to the bound potential in various forms, such as one-dimensional harmonic trap or other type of traps, repulsive or attractive potential of the δ-function, etc. These extensions are not trivial, some of them are quite difficult, and they are all closely related to the research of cold atom physics.

After returning to Tsinghua, Prof. Yang has published 13 articles in physics research. Among them, many have Prof. Yang as the sole author and some are collaborations of Yang with others, in which generally there is only one collaborator, mainly Ma Zhong-qi. Some other collaborators have done numerical calculations for Prof. Yang, like Wei B. B. of the Chinese University of Hong Kong. This shows that most theoretical research are mainly carried out and derived by Prof. Yang. This is unlike many others, who after becoming boss at 40 or 50 years old, no longer devote themselves to the frontlines of research.

In June of 2009, Yang once submitted a manuscript to *Physical Review Letters* (PRL), which turned out to be for him an unpleasant experience. The first reviewer dealt with the paper in a very sloppy manner, considered "all the results has been previously (in the sixties) obtained by the name-sake of the author," and completely ignored this manuscript presented the solution for a 1D problem *with a trap.*

The second reviewer's comments, "lacking of generality" and "lack of physical content," are in the category of criticism that lacks substantial content and many often encounter. Moreover, the editor's routine answer of "evidently an arrogant bureaucratic form letter" made Chen Ning Yang, one of founders of PRL and author of many important papers, feel that the whole process of rejection was "funny and troubling." Therefore, in the notes of *Selected Papers II With Commentaries*, Prof. Yang attached the reviewers' reports and two rounds of correspondence with the PRL editor, in an effort to address and change such situation.[4] Subsequently, his research articles were mainly submitted to *Chinese Physics Letters* (CPL), which is a journal of the Chinese Physical Society. This expressed Prof. Yang's point of view that the value of an academic paper is not equivalent to the impact factor of the journal it appears in. As the Editor-in-Chief of CPL, I often receive Yang's emails at midnight. As a senior scientist, Prof. Yang's energy and vitality is really admirable and amazing!

3. An Educator Committed to the Cultivation of Outstanding Talent

In his poem *Root*, there are two lines: "For those students who have great aspirations, I am willing to be the beacon which gives direction (for the path ahead)." Training China's gifted students is Prof. Yang's most-valued mission after his return, as well as the task he spends most time and effort on. As a Nobel Laureate in Physics, a master of both science and the arts as well as an educator well-versed in both Chinese and Western cultures, Prof. Yang has trained a multi-faceted, comprehensive pool of talented individuals, including undergraduates, doctoral candidates, postdocs and young visiting scholars. The training modes include the teaching of basic courses, lectures, discussions. In addition, he also tried to compare and contrast education systems in China and the United States in order to explore the modes for cultivating exceptionally talented undergraduates. The institutions involved include the Institute for Advanced Study, the Department of Physics in Tsinghua University, as well as dozens of colleges and universities nationwide.

Undergraduate education is the foundation of higher education. Teachers' basic responsibility is teaching. However, teaching is "a matter of conscience," as whether teachers try their best or not is hard for outsiders to assess. In the fall semester of 2004, regarding that many well-known professors in Chinese top universities did not teach undergraduate classes, Prof. Yang offered a semester of "College Physics" for four classes of more than 100 freshmen in Tsinghua's Department of Physics (Fig. 5). The 82-year-old Nobel laureate, with his unique insights and profound understanding, taught the basic physics course twice a week, two periods spanning 45 minutes each. And he often spent the five-minute breaks in continuing the spread of knowledge. Not only did the students benefit from his teaching, teachers from many colleges and universities also watched the live video in other classrooms as well, savoring Prof. Yang's distinctive teaching taste, styles, and subtleties.

Fig. 5. Prof. Yang is giving a class for first-year undergraduates. The left picture shows he is projecting his written content onto the screen. The picture on the right shows Prof. Yang surrounded by students who asked questions after class.

Prof. Yang, by his teaching freshmen, set a high standard for all university teachers and encouraged the vast majority of teachers who are active on the frontlines of teaching. His decision to teach also prompted many Tsinghua University teachers who have done well in research to step up to the teaching platform.

The Department of Physics in Tsinghua University offers two special programs to cultivate gifted undergraduates, one is the basic science class, and the other is the Tsinghua Xuetang (College) Physics Program. Both aim to explore how to cultivate gifted students and transform them into world-class scientists. Prof. Yang is very concerned about students' development in these two programs. Figure 6 is a photo taken on April 1st, 2010 when Chen Ning Yang gave a speech at the opening

Fig. 6. On April 1st, 2010, Chen Ning Yang is giving a speech at the opening ceremony of the Tsinghua Physics Program.

Fig. 7. Prof. Chen Ning Yang was listening carefully to a student's question when he had face-to-face discussion with students in the Physics Program.

ceremony of the Tsinghua Xuetang Physics Program. Prof. Yang said confidently with a raised finger that "10 years later, we will meet again to find out whether this education reform program is successful or not. I meant that seriously!" Time flies, the Tsinghua Xuetang Physics Program will soon celebrate its 10th anniversary, and then it is time for us to ask Prof. Yang to judge if this Program is a success or failure! I often invite Prof. Yang to discuss questions face-to-face with students in the Department of Physics and the Tsinghua Xuetang Physics Program, in my memory, he never refuses my invitations. Figure 7 is a picture of him listening carefully to a student's question when he has face-to-face discussion with students in the Physics Program.

The Institute for Advanced Study led by Prof. Yang has attracted, influenced and trained many young Chinese scientists through the excellent academic environments and superior mentors. Prof. Yang's lecture and article entitled "My Study and Research Experience"[7] have become Chen Ning Yang's "Ten Commandments" for students and young scientists. In recent years, Tsinghua's Department of Physics has cultivated a great number of graduates who have emerged in the international arena. Take for an example, the Sloan Research Fellowships, which was set up by the Sloan Foundation in the United States in 1955, aims to award outstanding young scholars in the early stages of their careers, and 43 of the winners have won the Nobel Prize while 16 were awarded Fields Medals. Since 2010, 8 graduate students of the Tsinghua Department of Physics have won Sloan Research Fellowships, including Xiao-liang Qi, Cen-ke Xu, Shi-na Tan, Xi Chen, Yue Shen, Dengke Ma, Lei Qi, and Xie Chen. Many of them benefited from the academic halls of the Science Building

Fig. 8. In 2011, Chen Ning Yang attended Ph.D. graduation ceremony of the Institute for Advanced Study.

and learned a lot in their undergraduate years. The Institute for Advanced Study of Tsinghua University has seen a total of 65 PhDs, most of them have cut a splendid figure. At least 10 "made-in-China doctors" work in foreign elite schools as professors, and much more work in domestic universities with various titles. For example, Dr. Xiaoliang Qi, who graduated in 2007, has already been promoted a full professor in Stanford University. Dr. Hui Zhai, who graduated in 2005 and was Prof. Yang's only doctorate candidate in Tsinghua, is now a full professor at IASTU and awarded "The National Science Fund for Distinguished Young Scholar" of NSFC. Dr. Zhai works closely with a lot of experimental groups, and is considered to be one of the best theoretical physicists in cold atom research field. Figure 8 is a picture of Ph.D. graduation ceremony shot in 2011 of the Institute for Advanced Study, Tsinghua University.

As a wise physicist and an educator who is very familiar with both China's and the United States' education philosophy and systems, Prof. Yang often compares the advantages and disadvantages of the two different education systems. Many of Prof. Yang's views, especially those on how to cultivate China's top scientific and technological personnel, deserve deep consideration. He has a clear understanding of the strengths and weaknesses of both education systems. He believes that Chinese university education benefits students who score 70–85, while American university education is good for those with score 90 or more. At a Sino-American symposium on the physics education, he said: "There is a profound difference between the education philosophy in China and in America for deep historical and cultural reasons. The word 'educate' can be derived from a Latin word which means 'to rear' and 'to raise.' But in Chinese, it is 'Jiao Yu,' two Chinese characters. "Yu" means 'to rear' that proceeds by 'Jiao,' 'to teach.' Indeed, that difference is significant, because in the Chinese philosophy of education, to teach is at least as important as to raise. And I think this profound difference has not been sufficiently analysed by educators, ministers and professors."

4. A Science Master as well as Unique Researcher on Science History

Among 27 SCI papers and 2 monographs published by Prof. Yang in Tsinghua, more than half is involved the study on the history of science and physics, his annotation on the evolution of some important concepts and physics theories, as well as the commentary on several physics masters. Prof. Chen Ning Yang's research on science history and science masters is very unique. Namely, many great masters of physics and mathematics he studied and commented on are almost of his generation, or slightly older than him. And he had been in direct contact with all of them to certain extent. In particular, the first 10 years of this century coincided with the centennial birthday of many physics masters, and he was invited to give talk in several international symposia devoted to the master and to write articles. Another unique feature of Prof. Yang's investigation on physics masters stems from the fact that most people in this field are not doing physics research by themselves, or are relatively junior physicists, thus often study the history of science from a perspective that often includes admiration for masters leading to some distortion in their research. However, Prof. Yang himself is a leading physics master in the second half of the 20th century so that his view is one of a horizontal perspective, with even a downwards slant at times. To be honest, Prof. Yang's accomplishments are nothing less than the masters he studied, and even are higher than some master. Therefore, he tends to stand in a higher position when viewing these masters, which makes his research original, colorful and precious. In addition, Prof. Yang's frankness and open-mindedness, his amazing memory and clear, detailed personal files make his research on the history of physics and science more valuable. These features make Chen Ning Yang's research articles on physicists and physics history extraordinarily precious. And I think today few people in the world can write such articles.

Prof. Yang's most important achievements in his study of science history are impressive due to the following aspects. Prof. Chen Ning Yang once summed up three main themes in the history of 20th century theoretical physics: quantization, symmetry and phase factor, all of which are very profound. During the commemoration of the World Year of Physics (2005) which celebrated Einstein's contributions to science, Prof. Yang attributed Einstein's success to his vision and opportunity. Yang believes that unlike Lorentz and Poincare, Einstein's free perception, i.e. a combination of far-forward vision and close observation, led to the special theory of relativity. In his opinion, Einstein originally applied the basis of modern theoretical physics — the principle of symmetry-dominated interaction — to investigation, combined with the generalized coordinates invariance and the equivalence principle to create general relativity. Prof. Yang particularly admired Einstein's apartness, pursuit and profound vision, believing that these characteristics changed the course of development of basic physics. Prof. Yang's essays discuss Einstein's far-reaching impact on theoretical physics, thoroughly inspiring future generations. In his article "The conceptual origins of Maxwell's equations and gauge theory,"[8] Prof. Yang also researched a period of history through an examination of original articles by Maxwell. He studied carefully how gauge freedom developed and how it evolved into a symmetry principle supporting the Standard Model of particle physics.

While studying masters of science, Chen Ning Yang found that the majority of successful European and American physicists were aggressive and practiced one-upmanship, e.g. Oppenheimer, Taylor and Feynman, etc. In contrast, he himself preferred physicists like Fermi, Guangzhao Zhou and R. E. Mills. Prof. Yang put forward a question: Are the great achievements made in Europe and the United States linked to the aggressive characteristics of most European and American physicists? This is a very interesting question, which I call "Chen Ning Yang's Conjecture." If we link this to China's education, it prompts the question whether Confucian culture plays a positive or negative role, or if it is a double-edge sword in terms of the cultivation of innovative talents? This question is worthy of further study. Prof. Yang considered himself influenced by a strong Confucian tradition. When he was a 12-year-old, he spent summer vacation in the Science Building, and his father invited a top student in Tsinghua's Department of History, Zeliang Ding, to teach him "Mencius." Although among the "Four Books and Five Classics" Chen Ning Yang only seriously studied "Mencius," he believed Confucian culture has a profound influence on him. During the commemoration of the 60th anniversary of Yang–Mills Gauge Field Theory in 2015, Prof. Yang reviewed the controversial debates he had with Feynman and others over the prospect of physics in the early 1960s. He believes that his judgement is correct, and the reason why he is correct is because he was influenced by Confucian culture and practice: "Each day I examine myself in three approaches."

5. Other Contributions

After Chen Ning Yang's return, he gave numerous public lectures and wrote many articles around several other topic areas related to China, such as Chinese culture, the international relations, the development, the society, the art, the aesthetics, and the archaeology, etc. Due to limited space and time, I will only discuss two examples here.

In 2004, Prof. Yang gave a speech titled "The Influence of the Book of Changes (Yi Jing) on Chinese Culture," he thought that the "Book of Changes" influenced the thinking mode of Chinese people, and this influence was one of the important reasons why modern science did not sprout in China. In addressing concern surrounding the question why modern science did not take hold in China, Prof. Yang expressed his belief that it could be attributed to the following reasons. The first is that traditional China did not put emphasis on the spiritual world, but on the practical world, and it did not pay attention to the abstract theoretical framework. The second is the imperial examination system. The third is the lack of emphasis on technology, which was considered "miraculous trickery." The fourth is the fact that traditional Chinese culture only has the induction but not the deduction, this meant that without a combination of both aspects, modern science could not occur in China. The fifth is the concept of harmony between man and nature. Among them, the fourth and the fifth are closely related to the "Book of Changes." Prof. Yang's views were unique. Although this has resulted in reactions that involved many different opinions, it was obvious that such discussion and debate are beneficial for deeper understanding of Chinese culture and the development of modern science in China.

Prof. Yang also made clear his opinions on issues relating to academic integrity in China's academic circles. On June 14th, 2010, at 11 o'clock at night, I received an email from Chen Ning Yang. This email was about a manuscript reviewed in *Chinese Physics Letters.* A doctoral student at Tsinghua Institute for Advanced Study, Mr. XX, submitted an article to the CPL, which had his sole authorship. The Assistant to Editor invited two reviewers to evaluate it. One referee gave a completely positive opinion for the manuscript, while the other asked the author to cite another three articles that seemed to be unrelated to the content of the manuscript. The author asked Prof. Yang for advice. Prof. Yang's judgment was that the second reviewer's request for XX to quote the three articles that "have absolutely nothing to do with XX's work," was a case of reviewers abusing their privileges for personal benefit. We did a quick investigation and resolved the problem. This case is related with the academic misconduct that occurs frequently in academia, and many are so accustomed to it that they turn a blind eye to it. Prof. Yang's "meddlesome move" represents a clear-cut stand against academic misconduct and is deserving of admiration.

After Prof. Chen Ning Yang's return, he began his second spring as a master of theoretical physics, and it has been very successful. He made great contributions to the academic community and played an outstanding role in advancing China's science and technology. We wish Prof. Yang good health, and look forward to celebrating his 100th birthday and his tea birthday (108 years old)!

Acknowledgment

I would like to acknowledge the support from the National Natural Science Foundation of China (Grant No. 11374173).

References

1. B.-F. Zhu, *Physics* **45**(10), 621 (2016).
2. B.-F. Zhu, *Physics* **43**(4), 276 (2014).
3. C. N. Yang and F. Weng, 曙光集 *Dawn Set* (Beijing Sanlian Bookstore, 2008).
4. C. N. Yang, *Selected Papers II With Commentaries* (World Scientific, Singapore, 2013).

5. B.-F. Zhu, *Physics* **38**(8), 575 (2009).
6. C. N. Yang, *Selected Papers I With Commentaries* (W. H. Freeman, New York, 1983), p. 77.
7. C. N. Yang, *Physics* **41**(1), 1 (2012).
8. C. N. Yang, *Physics* **43**(12), 780 (2014) [original in English: published in *Physics Today* **November 2014**, 45–51 (2014)].

https://doi.org/10.1142/9789819826636_0003

JOURNEY THROUGH STATISTICAL MECHANICS

C. N. YANG
Physics Department, State University of New York at Stony Brook
Stony Brook, NY 11794–3800, USA

Received 12 October 1988

My first involvement with statistical mechanics and the many body problem was when I was a student at The National Southwest Associated University in Kunming during the war. At that time Professor Wang Zhu-Xi had just come back from Cambridge, England, where he was a student of Fowler, and his thesis was on phase transitions, a hot topic at that time, and still a very hot topic today.

In the mid 30's there was already great interest in phase transitions. It came about mostly from metallurgy. People found that the specific heat of alloys as a function of temperature showed characteristic peaks. In particular, for example, people found that beta-brass, or copper-zinc alloys, could have very sharp peaks in its specific heat as a function of temperature. At that time the dominant idea concerning such phenomena could be described as mean field theory. The most famous mean field theory was Bragg and Williams' theory. Bragg and Williams were both Englishmen and England was very active in this area of research. That was the reason why Professor Wang Zhu-Xi and his contemporary Professor Zhang Zhong-Sui in Cambridge worked on their Ph. D. theses in this area. (Many people in the audience probably know about Zhang Zhong-Sui, a most brilliant physicist.)

But it was very clear soon that the mean field theory is rich in physical ideas but defective in mathematical precision. Joseph Mayer, who was a young researcher at Columbia at that time, explored this problem and developed a theory later called Mayer's theory of condensation which caused great excitement. At an international conference in the Netherlands in 1937 it was debated whether Mayer's theory was the right theory or not.

Based on 19th century's 'ideas about phase transitions in thermodynamics, it had been generally believed that in such transitions as vapor-liquid transitions, there were the thermodynamic functions of the vapor phase, and the thermodynamic functions of the liquid phase. Vapour and liquid were *different* dynamical systems and should be treated separately. Mayer's starting point was different. He had *one single* dynamic system, out of which both phases resulted. Physically this was against thermodynamic intuition. Mathematically it was against the perception that out of *one* partition function there cannot result *two* different thermodynamics. Born and Fuch later wrote about the discussions at the 1937 conference:

* This paper was based on an invited talk given at the Nankai Institute of Mathematics, 1987.

Reproduced from International Journal of Modern Physics B, Vol. 02, No. 06, pp. 1325–1329 (1988), DOI: 10.1142/S0217979288001177

"Mayer's theory . . . treats all possible molecular arrangements with their proper weight, as if there were only one phase. How can the gas molecules "know" when they have to coagulate to form a liquid or solid? (Proc. Roy. Soc. London A166, 391 (1938))

I learned about Mayer's theory through the following circumstance. Immediately after the 1937 conference Born-Fuchs and Kahn-Uhlenbeck wrote famous papers to analyse the validity or non-validity of Mayer's theory. Around 1940 or maybe 1941, Professor Wang Zhu-Xi gave a series of lectures on these papers and as an undergraduate, I participated in these lectures. They were beyond my level of knowledge at that time. However attending those lectures was decisive for my career later, because through the lectures I learned of the importance of the theory of phase transitions.

In 1945, I went to the United States, and became a graduate student in Chicago. Nobody in Chicago was really interested that much in phase transitions and that included J. Mayer himself. But because of my early fascination in China about phase transitions, I made several attempts to get into the field.

One attempt was to understand the 1944 paper of Onsager which had brilliantly and unexpectedly solved the two-dimensional Ising model. The paper was very difficult because it did not describe the strategy of the solution. It just described the detailed steps, one by one. Reading it I felt I was being led by the nose around and around until suddenly the result dropped out. It was a frustrating and discouraging experience. However, in retrospect, the effort was not wasted. Not at all, as we shall see.

Another attempt was to understand ferromagnetism. Here I started with F. Block's important papers on spin waves, which led to Bethe's 1931 paper on what is now known as Bethe's hypothesis, and Hulthen's 1938 paper which developed Bethe's method. Bethe's and Hulthen's papers resulted in transcendental equations and I got sidetracked into looking for *complex* solutions of these equations and was thoroughly confused. Again, this was a frustrating and discouraging experience. But again, in retrospect, the effort was not wasted.

My main research direction in Chicago was on particle physics. In the fall of 1949 I went to Princeton, to the Institute for Advanced Study, where I was to stay for 17 years, 1949–1966. At the Institute no one was working on statistical mechanics in 1949–1950. But it happened that one day in a station-wagon ride from Palmer Square at the centre of Princeton to the Institute, a ride of about 15 minutes, J. M. Luttinger, a fellow postdoc at the Institute, described to me a new paper of Bruria Kaufman. It had greatly simplified the algebra of Onsager's method, and presented a clear and simple strategy for the whole idea. Being familar with Onsager's difficult maneuvers, I quickly grapsed the new perspective. After arriving at the Institute, I worked out the essential steps and was very happy at finally gaining an overall understanding of Onsager's method.[1]

Some months later, having finished some projects in particle physics, I returned to Onsager's solution and realized that much more results could be obtained starting with his method. In January 1951 I began to work in this area seriously. What happened after this was described by me in 1982 as follows[2]:

I was thus led to a long calculation, the longest in my career. Full of local, tactical tricks, the calculation proceeded by twists and turns. There were many obstructions. But always, after a few days, a new trick was somehow found that pointed to a new path. The trouble was that I soon felt I was in a maze and was not sure whether in fact, after so many turns, I was anywhere nearer the goal than when I began. This kind of strategic overview was very depressing, and several times I almost gave up. But each time something drew me back, usually a new tactical trick that brightened the scene, even though only locally.

Finally, after about six months of work off and on, all the pieces suddenly fitted together, producing miraculous cancellations, and I was staring at the amazingly simple final reuslt . . . .

The result was the spontaneous magnetization of the Ising model which was my first piece of work in rigorous statistical mechanics. So the efforts in Chicago were not wasted—they had prepared me for absorbing rapidly what Luttinger explained to me in the station wagon.

A year later I suggested to C. H. Chang to generalize my calculation to the rectangular Ising lattice. His paper[3] contained what was probably the earliest speculation about the "universality" of critical exponents.

In the Fall of 1951, T. D. Lee came to the Institute. He and I had been close friends in Chicago[4]. After his Ph. D. in Chicago he had spent one year (1950–1951) at Berkeley where he was not happy. So I suggested to Oppenheimer, the Director of the Institute, to invite him to be a postdoc at the Institute. After he came we began the next natural project, to calculate the susceptibility of the Ising model. To compute the magnetization I had used a first perturbation. To compute the susceptibility we used a second perturbation. Unfortunately this led to a formula which was an order of magnitude more difficult to evaluate than the corresponding formula in the first perturbation. After a few weeks of struggle we gave up on the problem and turned our attention to the lattice gas, which was mathematically the same as the Ising model. In particular, Onsager's result and my result together can be translated into the *exact* boundary of the two-phase region in the p-V diagram for the lattice gas. The isotherms are flat in this two-phase region, and become curved in both the liquid and gas phases. But then why did Mayer's theory lead to isotherms that stayed flat into the liquid region? In tracking this question down we introduced the concept of the zeros of the partition function, with which great clarification resulted concerning the phenomena of phase transitions. A particularly elegant step in this research was the unit circle theorem which I[5] "fondly consider . . . a minor gem."

In retrospect, my exposure to Wang's lectures in China had prepared me for this work on phase transitions. I suppose for most research workers the same pattern occurs: what one is exposed to in undergraduate and graduate years are likely to be what one works on in later life.

During 1952–1953 de Boer of Amsterdam was visting the Institute. He had made important contributions to the theory of liquid helium. I asked him to give a series of lectures on this subject and got quite interested in the fascinating phenomena of the superfluidity of liquid helim. In particular, I tried to understand the relationship between Bose-

Einstein condensation and superfluidity. Nothing useful came out of this effort at that time, but a few years later, when Kerson Huang came to the Institute in 1955 and introduced me to the pseudopotential method of Fermi and Breit, my interest in liquid helium was rekindled. Huang, Luttinger and later T. D. Lee and I collaborated during 1955–1959 in several explorations of properties of Bose systems.

In April 1961 I went to Stanford University for a visit of several months. W. M. Fairbank was starting on his experiment with B. S. Deaver on magnetic flux quantization in superconducting rings and the first question he raised when I visited his laboratory was, if they do find flux quantization in superconducting rings, would that be a new physical principle (a new property of electromagnetism), or would that be a result of known physical principles. He raised this question because the earlier proposal of Onsager and of London on flux quantization had been shrouded in ambiguities on this point.[6] I thought about Fairbank's question and started with Nina Byers to analyse the free energy of superconductors in the presence of magnetic fields. When Deaver and Fairbank's experiment was successfully concluded, Byers and I have also reached the conclusion that their beautiful result did not indicate a new property of electromagnetism, but followed from usual quantum statistical mechanics. That was the first true understanding of flux quantization.

This work led me to try to understand the precise meaning of Cooper pairing and of BCS theory. In other words, I wanted to understand the precise meaning of the Bose condensation of Fermion pairs which was the basic explanation of the phenomenon of superconductivity. The result of this investigation was a paper entitled "Concept of Off-Diagonal Long-Range Order and the Quantum Phases of Liquid He and of Superconductors" which I am very fond of.

My next phase of work in statistical mechanics started with an attempt to find examples of dynamical systems which exhibit off diagonal long range order. This attempt was not particularly successful, but I was led back to the Bethe-Hulthen work that I had studied in Chicago. With my brother C. P. Yang I found that what had been frustrating in Chicago, the problem of complex solutions of the crucial set of transcendental equations, can be avoided by a very simple trick: replace $\cot^{-1} f$ by $\pi/2 - \tan^{-1} f$! This simple trick[7] allowed us to use continuity arguments to keep track of solutions and to systematize the mathematics. Once more, my Chicago efforts were not wasted at all, though at the time it had seemed so.

The assumption of Bethe in his 1931 paper turned out to have much more general applicability than the original use, and C. P. Yang and I had in our series of papers in 1966 decided to honor Bethe's insight by calling his assumption "Bethe's hypothesis".

One of the problems that can be attacked with Bethe's hypothesis was the one-dimensional delta function problem with bosons which had been solved by Lieb and Lininger and by McGuire in 1963–1964. In trying to solve the corresponding fermion problem in 1967 I tackled the more general Boltzmann case and showed that even for this very general problem, Bethe's hypothesis still works because a set of cubic equations in certain matrices $Y_{ij}{}^{ab}$ are satisfied:

$$Y_{jk}{}^{ab} Y_{ik}{}^{bc} Y_{ij}{}^{ab} = Y_{ij}{}^{bc} Y_{ik}{}^{ab} Y_{jk}{}^{bc} .$$

This same set of equations was obtained by Baxter in 1972 in an entirely different problem in statistical mechanics and has since been named the Yang-Baxter equation. Many physicists and mathematicians have in recent years studied various solutions of these equations as well as their applications. Among these there are the especially important work of Jones and others that exploit the similarity of the Yang-Baxter equation with the fundamental equation of the braid group:

$$ABA = BAB$$

to obtain new invariants in the theory of knots. This and other applications of the Yang-Baxter equation seem to indicate that these cubic equations embody some fundamental structure still to be explored.

References

1. C. N. Yang, *Selected Papers (1945–1980) with Commentary* (hereafter referred to as SP), p. 12, (Freeman, 1983).
2. SP, p. 12.
3. C. H. Chang, *Phys. Rev.* **88** (1952) 1422.
4. SP, p. 7, pp. 53–54.
5. SP, p. 15.
6. C. N. Yang, Talk at the 1982 Conference to celebrate W. M. Fairbank's 65th Birthday, (to be published). London's argument, if correct, would lead to the conclusion that all fluxes are quantized, and would have nothing to do with superconductivity, which is of course incorrect. Onsager appeared to have believed that flux quantization, if found, is somehow a fundamental property of electromagnetism, like Dirac's magnetic *monopole* (which is independent of superconductivity).
7. For more details, see *Phys. Rev.* **150** (1966) 321, footnote 13.

https://doi.org/10.1142/9789819826636_0004

MY EXPERIENCE AS STUDENT AND RESEARCHER

C. N. YANG
Tsinghua University, Beijing, China
Chinese University of Hong Kong, Hong Kong

Received 14 March 2012
Accepted 15 March 2012
Published 3 April 2012

I was a student in Chongde (崇德) Middle School for four years 1933–1937, from Grade 7 to Grade 10. The school had about 300 students. It had a small library in which I developed the habit of browsing around. It was in that library that I had a first glimpse of modern physics through reading a Chinese translation of James Jeans' *The Mysterious Universe* (Fig. 1).

Fig. 1. Cover of the book *The Mysterious Universe* (Cambridge University Press).

Reproduced from International Journal of Modern Physics A, Vol. 27, No. 09, 1230009 (2012), DOI: 10.1142/S0217751X12300098

In the book Jeans used daily language to describe the development of special relativity in 1905, of general relativity in 1915 and quantum mechanics in 1925. I was fascinated.

In 1937 the war against Japanese invasion broke out. My family had to move through a long and difficult journey, arriving finally in the Spring of 1938 at Kunming (昆明). I still vividly remember how the train in which we were riding between Wuhan (武汉) and Guangzhou (广州) was bombed midway in Hunan Province (湖南). In Kunming I skipped Grade 12 and took the entrance examination to the Southwest Associated University (Lianda 联大).

I had not studied high school physics, so to prepare for the entrance examination I borrowed a copy of a standard high school physics textbook and read it through in several weeks. At the end of this reading I concluded that physics was the subject that I liked. Thus when I registered at Lianda I enrolled in the Department of Physics.

I still remember, that textbook stated that the acceleration in a uniform circular motion is centripetal and not along the tangential direction. I felt at first that this contradicted my intuition. But after thinking about it for a couple of days I finally realized that *velocity is a vector*. It has not only magnitude but also direction. This story taught me an important lesson: We all have intuitions. Most of them are correct. But some need revision. Thus

> **On the one hand our intuitions are extremely important. But on the other hand one must constantly absorb new concepts to revise one's intuitions.** (1)

In the Spring of 1942, to write a B.Sc. thesis required by the University, I went to see Professor T. Y. Wu (吴大猷) (Fig. 2), asking him to be my advisor. Forty years later I thus described how Professor Wu gave me guidance.[1]

Fig. 2. Professor T. Y. Wu (1907–2000) and I (taken in Stony Brook in 1982).

> He gave me a copy of an article by J. E. Rosenthal and G. M. Murphy in the 1936 volume of *Reviews of Modern Physics.* It was a review paper on group theory and molecular spectra. I was thus introduced to group theory in physics. In retrospect I am deeply grateful to Wu for this introduction, since it had a profound effect on my subsequent development as a physicist.

Group theory's applications in physics later became my principal area of work, comprising approximately two thirds of my research efforts.

In the fall of 1942 I enrolled as a graduate student in the Physics Department of Lianda. My M.Sc. thesis advisor was Professor J. S. Wang (王竹溪) (Fig. 3). His specialty was statistical mechanics and he guided me into this area of research. Approximately one third of my later research work was in this area.

Fig. 3. Professor J. S. Wang (1911–1983) (taken by H. T. Nieh in 1980).

Many years later I made the following description of my graduate days in Lianda:[2]

> During the academic year 1941–42 I was a senior in the Physics Department at the National Southwest Associated University in Kunming. The Department was quite small, with about ten faculty members, ten instructors, a few graduate students and not more than 20 students in each undergraduate class. When the academic year started in the fall of 1941, a new face appeared, auditing many of the senior and graduate courses and participating in all discussions. That was Huang Kun. He had already received his bachelor's degree in physics from Yenching University in Beiping, and had come to Kunming to join the Southwest Associated University as an instructor. Soon we got to know each other well, and that was the beginning of half of a century of warm friendship.

Fig. 4. Huang Kun (黄昆), Sheldon Chang (張守廉), and I (taken at the birthday party for celebrating Zhou Pei-Yuan's birthday at Peking University on June 1st, 1992).

Two of the courses we took that year were Professor Ta-You Wu's (吴大猷) classical mechanics and quantum mechanics. Quantum mechanics was a revolutionary development in physics which had started in Germany, Switzerland, England and Denmark in the years 1925 to 1927. Professor Wu was the physicist in China who trained during the 1930's and early 1940's the greatest number of students in this important new fundamental branch of physics. I remember many discussions after Professor Wu's lectures. It was through these discussions that I first got to know Huang Kun well, as a physicist and as a person.

A year later, in the summer of 1942, Huang Kun and I both enrolled as graduate students at the Southwest Associated University. He worked on his thesis with Professor Wu on atomic and molecular problems in astrophysics, while I worked on mine with Professor Wang Zhu-Xi (J. S. Wang) (王竹溪) on statistics mechanics. The graduate students' stipend was meager and we all looked for teaching positions to supplement our incomes. My father was a friend of President Xu Ji-Zu (徐继祖) of Kun Hua Middle School (昆华中学) who arranged to have Huang Kun, me and Zhang Shou-Lien (张守廉 Sheldon Chang), a fellow physics graduate student, split a teaching position at the School. As part of the package deal, the School gave us a corner room in one of their new buildings for the three of us to live in.

The School was situated approximately 3 km from the campus of the Southwest Associated University. Huang, Chang and I would spend our day on the University campus, taking our meals in the mess hall on campus, returning to our room at the School in the evening to sleep. There was no facility for producing potable water on the University campus, so we developed the habit of spending an hour or two everyday after supper, drinking tea in one of the tea houses clustered along the three streets near the campus before returning to the School. During these hours and hours of tea drinking, we really got to know each other well. We discussed and argued about everything under the sun: from ancient history to contemporary politics, from large cultural patterns in the world to small details in some movies we had recently seen. In all of these, I remember Huang Kun as a fair debator, not given to tricking his opponents. I also remember that he had a tendency to push his arguments to the extreme. Many years later when I reflected on this, I found it interesting how this tendency of his had seemed to be totally absent in his physics.

Our fellow tea drinkers formed a colorful and curious group. There were many students like us. But the majority of the tea drinkers were towns folks, horse carriage drivers, traders from distant counties and the like. Everyone was noisy. We especially. Oftentimes during a heated debate we would suddenly become aware that we were by far the noisiest and that everyone was watching us. (That awareness may or may not terminate our argument.) However, there was rarely any animosity between us and the nonstudents in the tea houses.

We experienced, during these hours of tea drinking, scenes and events that I was never to forget. Several times, sitting in the tea houses along Fung Du Street (凤翥街) which led northward away from the city into small hills dotted with simple tombs, we saw processions of soldiers marching in the northward direction with one or several prisoners in their midst. Each prisoner had a white picket-like cardboard tied to his back with his name and crime written on it. Most of the prisoners marched in silence. A few would yell something which sounded like: "Twenty years later, again a brave fellow!" Every time such a procession passed, the noise level in the tea house would drop perceptibly. Then the expected distant gun shots, and we would sit silently, waiting to watch the soldiers marching back.

Against such background happenings, we argued endlessly about physics. I remember once our topic of discussion was about the meaning of measurements according to the Copenhagen interpretation of quantum mechanics. That is a very subtle subject, and our arguments started at tea, lasted throughout the evening, and continued back in our room at the Kun Hua Middle School. After the electric lights were turned off and we were all in bed, the arguments did not stop.

It is no longer clear to me today what precise point we were arguing about that evening, nor who among us took which side. But I vividly remember that all three of us eventually got up from our beds, lit candles and examined in detail a few paragraphs of Heisenberg's *The Physical Principle of the Quantum Theory* to settle our argument.

Huang was an avid reader of English novels. It was he who introduced me to Joseph Conrad, Rudyard Kipling, John Galsworthy and other authors. Many of the novels of these authors we borrowed from the University libraries. Others we bought from the stalls that sold K-rations, army boots, cans of cheese as well as used pocket books with which the U.S. military personnel flooded the flea markets of Kunming.

It was a simple life we led in those years, as I have described elsewhere. A dish of peanuts to go with the tea was a great treat, which we could not indulge in very often. It was not an austere life, for we did not expect and did not desire more material things. It was also not a frustrated life, for we found plenty of intellectual stimulation and satisfaction. It was just a very simple life which, without our realizing it at the time, had shaped our tastes and styles in physics in ways that were to have profound effects on our later careers.

My discussions/arguments with Huang Kun and Sheldon Chang and my later experience as a professor of physics taught me:

Discussions with classmates offer opportunities for deep understanding. (2)

Sheldon Chang later moved to Electrical Engineering and Control Theory, obtaining his Ph.D. in the U.S. He is now a Professor Emeritus of Stony Brook University. Huang Kun (1919–2005) later received his Ph.D. degree in Great Britain, specializing in solid state physics in which he made very important contributions. He was also the father of semiconductor research in China. In 2001 he received the highest Science Award of China.

A few weeks after the end of World War II in 1945 I flew on a DC3 to Calcutta, where I had to wait for a berth on ships bound for the U.S. for several months, finally arriving in New York on November 24th after a long voyage through the Red Sea, Suez Canal and the Mediterranean. In January 1946 I enrolled as a graduate student in the Physics Department of the University of Chicago. My aim was to write an experimental Ph.D. thesis with Professor Enrico Fermi (1901–1954, Fig. 5).

In 2001 there were celebrations in Chicago and in Rome for Fermi's 100th birthday. In the paper I read at these celebrations was the following passage:[3]

> Enrico Fermi was, of all the great physicists of the 20th century, among the most respected and admired. He was respected and admired because of his contributions to both theoretical and experimental physics, because of his leadership in discovering for mankind a powerful new source of energy, and above all, because of his personal character. He was always reliable and trustworthy. He had both of his feet on the ground all the time. He had great strength, but never threw his weight around. He did not play to the gallery. He did not practise one-up-manship. He exemplified, I always believe, the perfect Confucian gentleman.

Fig. 5. Enrico Fermi (taken in 1940's).

Fig. 6. Edward Teller and I (taken in 1982).

At the time the Physics Department of the University of Chicago was ranked No. 1 or 2 in the world. Besides Fermi, another important theoretical professor was Edward Teller (1908–2003, Fig. 6). He was a brilliant scientist who had made important contributions in physics and in chemistry. Later in the early 1950's he discovered the method of constructing a hydrogen bomb and became world famous.

My plan to write an experimental thesis with Fermi did not materialize because his laboratory at the time was in the Argonne National Laboratory which was not open to me. So Fermi recommended that I first work with Teller on theory.

During the first half of 1946 I was Teller's graduate student. The first research problem he gave me was about the lifetimes of K-capture in Be and in BeO. He proposed that I use the Thomas–Fermi–Dirac and the Wigner–Seitz approximations to do the calculation. Several weeks later I showed him my results and he happily arranged for me to give a seminar. That was my first seminar report in America. In the small audience of less than twenty people there were several very important scientists: Fermi, Urey, the Mayers, etc. My report received universal approval, and Teller asked me to write it up for publication. I then spent approximately a whole week to do this but without success, since I could not adequately estimate the reliability of my results which involved several different kinds of approximations. Fortunately Teller was not offended and he gave me another problem on nuclear physics.

Teller had six or seven graduate students at that time. We met to discuss physics once or twice a week, oftentimes also to have lunch with him. Teller had lots of new ideas concerning nuclear physics, solid state physics, cosmic rays, etc. Gradually I found his style of research was different from my own. Thus while I continued to participate in his discussion group, I began to look for theoretical problems myself.

In the fall of 1946 Fermi recommended me to work in the laboratory of Professor Allison. Allison was an experimental nuclear physicist. At the time he was constructing a 400 keV Cockroft–Walton accelerator. He had five or six graduate students and accepted me as an additional one. In the meantime I continued to participate in Teller's discussion group.

At the time I was one of the most prominent graduate students in the Physics Department because what I had learned in Lianda of fundamental theoretical physics had already reached the frontiers of the field at the time. But I was very clumsy in the laboratory. Laboratory mates respected my theoretical knowledge, oftentimes asking me to help them to solve theoretical problems. But they laughed at my blunders in the laboratory. "Where there is a Bang, there is Yang!"

1947 was a year of unhappiness for me. In a letter I wrote to Huang Kun who was then a graduate student in England, I had used the word "disillusioned" to describe my feelings at the time. Why? Because although I was trying hard but I did not have the talents to do experiments. On the other hand in theoretical physics all the topics that I had concentrated on myself were leading to pure frustration.

It is very common for a graduate student to feel discouraged in looking for a good problem for his/her thesis work. (3)

The theoretical problems that I worked on myself in the year 1947 included the following items:

(1) Onsager's 1944 paper on Ising Model.
(2) Bethe's 1931 paper on Spin Waves.
(3) Pauli's 1941 review paper on Field Theory.
(4) Many papers after 1943 on angular distributions in nuclear reactions.

Of these four topics the first two were in statistical mechanics. I was interested in them because of the influence of Professor J. S. Wang. The other two problems were related to the concept of symmetry. I was interested in them because of the influence of Professor T. Y. Wu.

In the Physics Department in the University of Chicago at that time nobody was interested in the first three of these problems. I worked on them by myself in the department library, trying to understand the papers on these problems, and hoping to make further advances. The end result of several weeks of hard work on each of these three topics was total frustration.

Fortunately Teller was very much interested in the fourth problem. At that time there were many theoretical papers on this topic. I found all of them were intuitive and lacked rigour. So I spent several weeks applying the concept of "Invariance of

Physical Laws under Space Rotation" to substantiate the intuitive ideas floating in the literature on the subject. The effort was successful and I wrote a short paper which Teller very much liked. At that time everybody in the department knew that Frank Yang's work in Allison's laboratory was unsuccessful. One day in the spring of 1948 Teller came to Allison's laboratory[4] and suggested that I *abandon* my plan to write an experimental thesis. He would sponsor my short paper as my Ph.D. thesis after I had extended it to include relativistic cases. I was at first disheartened by Teller's proposal, but after a few days recovered and accepted his suggestion with relief.

Altogether I spent about 18 to 20 months in Allison's laboratory. Was that a complete waste of time? No, absolutely not! Through that experience I learned that the value judgement of an experimental physicist was very different from that of a theoretical physicist. This understantding influnenced many of my later research work, including the 1956 work on the possibility of parity nonconservation and my 1964 paper with T. T. Wu on phenomenological analysis of CP invariance.[5]

My Ph.D. thesis was the first paper I published concerning invariance and non-invariance of physical laws. Soon after I published a second paper in this field, about the spin of π^0 meson through a careful analysis of the group theoretical representation of invariance in field theory. These two papers established me as one of the foremost theorist on the use of group theory in analysing symmetry properties. It was then that this field was beginning and I was very fortunate to have gotten into it at its infancy.

It is best to enter a research area when it is new and developing. (4)

After I got my Ph.D. degree in the summer of 1948, the University of Chicago appointed me as an instructor. At that time the hottest theoretical area of research was renormalization theory. The three theory professors in Chicago: Fermi, Teller and Wentzel were not working in this field. Therefore after one year 1948–1949 I moved to the Institute for Advanced Study (IAS) in Princeton. Before leaving Chicago Fermi told me it was good to spend some time at the IAS, but the work there was too academic. It was a bit like a medieval cloister, he said, and was not a place for long stay. So he suggested that I go to the IAS for one year and then return to Chicago. I was of course totally in agreement with this advice of his. But[6] because of the convenience of dating Miss Tu (later my wife) in New York City, I did not return to Chicago after one year and instead remained at the IAS for altogether seventeen years, 1949–1966.

During these seventeen years the four topics that I had worked on by myself in Chicago, described above, all became successful areas of research for me. The breakthrough for the first item, Onsager's work on the Ising Model, had come about accidentally:[7]

> One day in early November, 1949, in a ride in the station wagon that the Institute ran from Palmer Square opposite Princeton Univeristy to the

> Institute, J. M. Luttinger and I happened to talk about the Ising model. Luttinger said that Bruria Kaufman had simplified Onsager's method so that the solution could be understood in terms of the representation of a system of $2n$ anticommuting Hermitian matrices. I knew such representations well and understood quite readily the main points of the Onsager–Kaufman method. After arriving at the Institute, I worked out the essential steps of this approach and was very happy at finally understanding Onsager's solution...
>
> I kept thinking about it, and realized that Onsager and Kaufman had obtained much more information than just the partition function,...
>
> I was thus led to a long calculation, the longest in my career. Full of local, tactical tricks, the calculation proceeded by twists and turns. There were many obstrctions. But always, after a few days, a new trick was somehow found that pointed to a new path. The trouble was that I soon felt I was in a maze and was not sure whether in fact, after so many turns, I was anywhere nearer the goal than when I began. This kind of strategic overview was very depressing, and several times I almost gave up. But each time something drew me back, usually a new tactical trick that brightened the scene, even though only locally.
>
> Finally, after about six months of work off and on, all the pieces suddenly fitted together, producing miraculous cancellation, and I was staring at the amazingly simple final result.

Why was I able to "understood quite readily the main points of the Onsager–Kaufman method"? One: I had *thoroughly* studied the representation of $2n$ anticommuting Hermitian matrices when I studied Dirac's equation in Kunming. Two: I had spent several weeks in 1947 in Chicago trying to understand Onsager's 1944 paper, without success. But that experience had *thoroughly familiarized me* with Onsager's paper. Combining these two earlier efforts I was able to rapidly understand what Luttinger had said about the real strategy of Onsager's solution.

What I had learned from Professor J. S. Wang in Kunming made me deeply *interested* in statistical mechanics, leading to my *preparatory efforts* in Chicago on Onsager's work. The final *breakthrough* came from a short conversation with Luttinger. This process:

$$\textbf{Interest} \rightarrow \textbf{Preparation} \rightarrow \textbf{Breakthrough}, \qquad (5)$$

I think, is the necessary three-step process for any research work. In the experience described above the breakthrough came from an external stimulation (conversation with Luttinger). But in most cases the breakthrough is a sudden *inspiration* without external stimulation: After long preparation, when one is not thinking about the problem, apparently new conceptual combinations were still being formed subconsciously in one's brain. When the right combination occurs a sudden *inspiration* would result. A century earlier Poincaré[8] had given a very interesting analysis of this process.

The third problem I had unsuccessfully worked on in Chicago was about the concept of gauge invariance in electromagnetism in Pauli's 1941 review paper. Gauge invariance was invented by Hermann Weyl (1885–1955, Fig. 7) in the years 1918–1929. I was deeply interested in this beautiful concept and wanted in 1947 to generalize it. (Why had my fellow graduate students at the time not launched into similar efforts? I think it is because I was especially interested in group theory and in the concept of invariance in physics, subjects in which most of my fellow graduate students at the time were not interested.)

Fig. 7. Hermann Weyl (1885–1955).

I started with an important formula in electromagnetism

$$F_{\mu\nu} = A_{\mu,\nu} - A_{\nu,\mu} \tag{A}$$

and generalized it to

$$F_{\mu\nu} = B_{\mu,\nu} - B_{\nu,\mu}\,, \tag{B}$$

in which B_μ is a 2×2 square matrix, unlike A_μ, which is a simple 1×1 matrix. This very natural generalization however led to increasingly complicated calculations. Thus I had to give up. That was in 1947. My aim was to use this generalized gauge invariance to construct the mutual interactions between the many particles, Λ, K, etc., which were newly discovered at the time. Figure 8 are reproductions from three pages of my notes of 1947.

More and more new particles were discovered in the succeeding few years. So several times I returned to this attempt at generalizing gauge invariance to formulate interactions between them. But each time the formulae became *increasingly complex*, increasingly ugly, and I had to give up. Finally in 1953 to 1954 I was

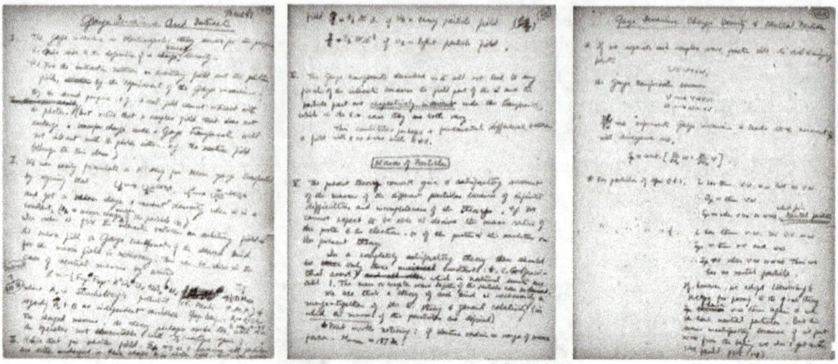

Fig. 8. Reproductions from three pages of the notes of 1947.

visiting the Brookhaven National Laboratory (BNL) for one year. I had two young officemates at BNL that year. One was Robert Mills (1927–1999, Fig. 9) who was a student of Norman Kroll (1922–2004) and was just about to get his Ph.D. degree. Another one was an experimental physics graduate student Burton Richter (1931–) who later shared the Nobel Prize in physics in 1976 with Samuel Ting.

Fig. 9. Robert Mills and I (taken in Stony Brook on May 22nd, 1999).

Naturally in my discussions with Mills I mentioned my unsuccessful attempts to generalize gauge invariance. One day somehow we said although formula (B) was quite natural but maybe it should be amended to become

$$F_{\mu\nu} = B_{\mu,\nu} - B_{\nu,\mu} + (B_\mu \text{ and } B_\nu \text{ polynomial})\,, \tag{C}$$

in order to possibly cancel out the "increasingly complex" terms. We decided to first try a quadratic polynomial. If that did not work we would try a cubic one, etc., etc. Fortunately we rapidly found that if we replaced (B) with

$$F_{\mu\nu} = B_{\mu,\nu} - B_{\nu,\mu} + B_\mu B_\nu - B_\nu B_\mu \,, \tag{D}$$

the subsequent calculation became *increasingly simple*. Thus we knew we had uncovered a great treasure!!! [We did not know then that from a geometrical viewpoint it was *natural* to include the quadratic terms in (D).]

With this breakthrough we followed the theory of Maxwell (1831–1879) and quickly wrote down a set of equations for generalized gauge theory. It was truly beautiful. But it had a problem: It seemed to imply that there should be massless charged particles which had never been seen. Besides it was difficult to understand how a charged particle could be massless. This problem caused us several months of complicated and fruitless research. There was also a famous episode of unpleasant criticism from Pauli.[9] Finally we decided although we did not have a good solution to this problem, the whole idea was too beautiful not to be published. Our paper was sent to the *Physical Review* in June, 1954. Fortunately it was immediately accepted and was published in October.

This paper is my most important contribution to physics. It led to the principle that *symmetry dictates interactions* which in a sense is one part of Einstein's *geometrization of physics.*

The paper did not answer the massless particle question mentioned above, but our decision to publish, in retrospect, was absolutely correct. I learned from this:

> **It is often not possible to solve at once all aspects of a difficult problem.** (6)

About the problem of massless particles, the idea of *symmetry breaking* was introduced around 1970. With this additional idea there later was developed a very successful "standard model." Since at that time I did not like to introduce symmetry breaking into a fundamental theory,[10] I lost the opportunity to make contributions in this area of research.

About the collaboration between Mills and me, Mr. Wang Zhi (王志) of CCTV interviewd me on January 26th, 2005 in Beijing. He asked why my research work was often done in collaboration with other physicists. The following was my answer:[11]

> There are many advantages in scientific collaboration. When you are working on a problem, sometimes you get stuck, you get discouraged. If at that time another person discusses the problem with you, asking you questions, suggesting a new direction for consideration, etc., your interest might be rekindled.

Thus I believe:

> **Discussions with colleagues is oftentimes a very fruitful method of research.** (7)

Between 1954 to 1956 many new particles were found. Particularly interesting were two particles θ and τ, which decay into two or three π mesons:

$$\theta \to \pi + \pi\,,$$

$$\tau \to \pi + \pi + \pi\,.$$

Increasingly more accurate experiments showed that θ and τ have approximately the same masses and the same lifetimes, indicating they might in fact be the same particle which had two different modes of decay. There is nothing strange about this EXCEPT IT WOULD VIOLATE THE LAW OF PARITY CONSERVATION: According to this law the "parity" of two π's is $+1$ and of three π's is -1. If θ and τ are the same particle then it can decay into $+1$ parity and also into -1 parity. Parity is thus not conserved. That is absolutely impossible! This dilemma was called the θ–τ puzzle and was the most puzzling problem in fundamental physics in the years 1954 to 1956. In a 1957 paper I said:[12]

> The situation that the physicist found himself in at that time has been likened to a man in a dark room groping for an outlet. He is aware of the fact that in some direction there must be a door which would lead him out of his predicament. But in which direction?

In the summer of 1956 T. D. Lee and I (Fig. 10) were trying to find this door. After carefully examined the five types of experiments which had been considered to have established parity conservation in β-decay, we found to our surprise that in fact *none of them had proved parity conservation*: all of them were *not complicated enough.* We also pointed out several kinds of *more complicated* experiments which could decide whether parity was conserved in weak interactions such as in β-decay.[a]

Fig. 10. T. D. Lee and I (taken at IAS in 1957).

[a]Weak interactions include the decay of θ, τ, etc. and also β-decay.

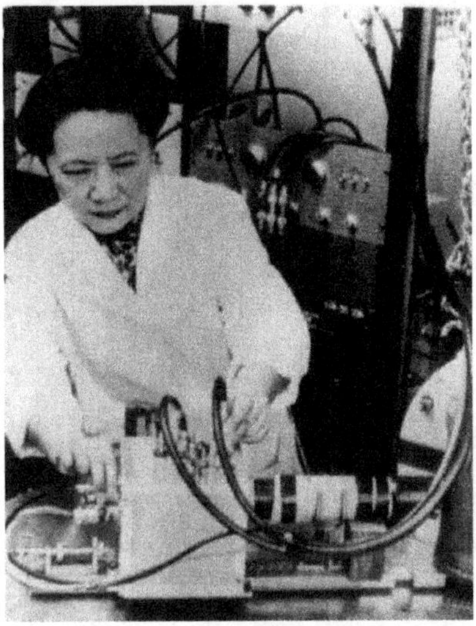

Fig. 11. C. S. Wu (1912–1997).

In June 1956 we wrote these results up and submitted it for publication in the *Physical Review.* We also sent copies of this preprint to many colleagues. Rapidly and uniformly came the response: Parity is definitely conserved. The experiments suggested by Yang and Lee were unnecessary and would only waste human and material resources. Fortunately C. S. Wu (1912–1997, Fig. 11) thought otherwise. She was influenced by Pauli and did not believe that parity could be nonconserved. But she thought since no past experiments had verified that parity was conserved in β-decay, it was important to do *an experiment to test this fundamental law of nature.* After six months of efforts she announced at the beginning of 1957 that in fact in β-decay *parity was not conserved*. Furthermore it was maximally nonconserved. Her announcement shocked the whole world of physics including: particle physics, nuclear physics, atomic and molecular physics. As to why the physical universe satisfies, on the one hand very accurate left–right symmetry (parity conservation) in most interactions, but on the other hand left–right asymmetry (parity nonconservation) in weak interactions, is a deep mystery still unsolved today.

That epoch-making success of Wu's experiment taught her:[13]

Never believe in laws which are considered self-evident requiring no experimental proof. (8)

The second problem I had tackled unsuccessfully in 1947, about Bethe's 1931 paper, became important for me later in the 1960's through a round about way. In the summer of 1961 I visited Stanford University. It happened that Fairbank and Deaver were doing their experiments on flux quantization in superconducting

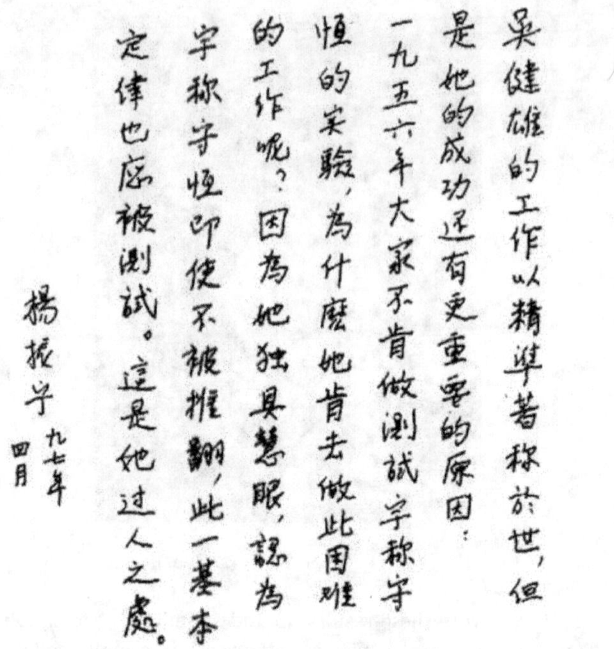

吴健雄的工作以精準著称於世，但是她的成功还有更重要的原因：一九五六年大家不肯做测試宇称守恒的实験，為什麽她肯去做此困难的工作呢？因為她独具慧眼，認為宇称守恒即使不被推翻，此一基本定律也應被測試。這是她过人之處。

楊振宁 九七年四月

Fig. 12. A paragraph I wrote after Wu passed away in 1997.

rings. Their experiments induced me to start working on superconductivity, which led later to the concept of ODLRO, which I have always been very fond of.

A couple of years later, in order to find a mathematical model which rigorously exhibit ODLRO, T. T. Wu, C. P. Yang and I tried many different models. In these considerations we returned to Bethe's 1931 paper. But this time we started to investigate a *generalized* Bethe's problem, thereby introducing the concept of *analytic continuation* to the model. With this generalization the complicated equations in Bethe's paper became controllable. Thus we were able to gain better understanding of their solutions, and between 1966–1969 C. P. Yang and I published several good papers based on this concept. They turned out to be my first research papers in Stony Brook.

In retrospect this experience also followed the three steps:

Interest → Preparation → Breakthrough,

but this time the breakthrough came about because of a new factor: To study the problem from a generalized angle. Thus:

Putting a problem in a generalized context is often a good strategy. (9)

As a matter of fact, my 1954 work with Mills was another example of the usefulness of generalization: We generalized the gauge invariance of electromagnetism to non-Abelian gauge invariance.

I moved from the IAS in Princeton to Stony Brook University in 1966. At Stony Brook I began to regularly teach undergraduate and graduate courses, which I did not do in the IAS. One day in 1967 or 1968 I was teaching a course on general relativity. As I wrote down on the blackboard the famous curvature equation of Riemann:

$$R^l_{ijk} = \frac{\partial}{\partial x^j}\left\{ \begin{matrix} l \\ ik \end{matrix} \right\} - \frac{\partial}{\partial x^k}\left\{ \begin{matrix} l \\ ij \end{matrix} \right\} + \left\{ \begin{matrix} m \\ ik \end{matrix} \right\}\left\{ \begin{matrix} l \\ mj \end{matrix} \right\} - \left\{ \begin{matrix} m \\ ij \end{matrix} \right\}\left\{ \begin{matrix} l \\ mk \end{matrix} \right\}, \qquad \text{(E)}$$

I noticed its similarity to the equation of gauge theory exhibited above as (D). After the lecture I examined these two equations in detail and realized they were in fact both special cases of a general type of equation. In great excitement I went to Jim Simons, who was then the Chair of Mathematics Department in Stony Brook. He told me they were both equations in *fiber bundle* theory and gave me a copy of a standard monograph on the subject by Steenrod. The book turned out to be too dry and too formal for me and I did not learn anything from it. A few years later I asked Simons to give us physicists informal lectures on the basic ideas of fiber bundle theory. These lectures were extremely helpful to us, enabling us to realize that the basic *physical* concepts of all gauge theory (including electromagnetism), such as *potentials*, *field strengths*, etc., are in fact identical to beautiful basic *geometrical* concepts in fiber bundle theory. *I was deeply awed and inspired by this realization.*

Two lessons I learned from this experience:

Fundamental physics is based on beautiful mathematics. (10)

But not all beautiful mathematics find their way into physics. (11)

In Stony Brook I began to have Ph.D. students. My style of guiding students was such that I rarely had more than one graduate student at any time. So altogether I graduated only about ten Ph.D. students. But I am proud to have influenced the career of several graduate students some of whom were in fact not my own: They had come to Stony Brook aiming at specializing in high energy physics. I told them high energy physics did make spectacular progress in the years before 1980, but after that it was difficult for a young person to make good contributions in that field. The big accelerations were becoming increasingly expensive, and the collaborative experiments were becoming too too big. Unfortunately many young people did not realize these and continue to enter the field, causing overcrowding, and the number of good ideas/person/year in the field became very small.

Several Stony Brook graduate students did take my advice seriously and I am happy today that they are now very successful in their respective fields different from high energy physics.

A graduate student had better not choose a field which is becoming overcrowded. (12)

Fig. 13. Painting by Fan Zeng in 2004.

I have listed above, in (1) to (12), personal experiences which might be helpful to a graduate student of physics. Of these, perhaps the one most worthy of attention is

$$\textbf{Interest} \rightarrow \textbf{Preparation} \rightarrow \textbf{Breakthrough.} \tag{13}$$

Two comments about this process:

(a) My father was a professor of mathematics. When I was a little boy he taught me traditional Chinese mathematical problems like “Chicken and Rabbits in a Cage” (鸡兔同笼), “Han Xin Counting Soldiers” (韩信点兵), etc. I learned quickly and he was pleased. Later I had three children in the U.S. When they were little I also taught them these problems. They all learned quickly and I was pleased. But there was a difference between me and my children: One year after learning how to solve these problems, I still remember them, while one year after learning these solutions, my children had completely forgotten having ever heard of them.
It seems that the SAVE and RETRIEVE systems in one’s brain are highly selective. And these systems are structured differently for different individuals. If for one individual these systems happen to be *partial* to one kind of information, that could be a seedling. With nourishment and sunlight and care the seedling could *slowly grow* and eventually may *flower*, completing the three-step process (5).

(b) Fan Zeng (范曾), the poet and painter, had presented in 2004 a big painting (Fig. 13) to the Chern Institute of Mathematics of Nankai University. He added to the painting a beautiful poem. Its last line describes how profound undauntable *love*, plus long *efforts*, plus *inspiration* produce literature. I do not remember Fan had ever discussed the creative process in science with either Professor Chern or with me. His poem seems to indicate that the process of creation for an artist follows also the same three steps as for us scientists.

References

1. C. N. Yang, *Selected Papers 1945–1980 with Commentary* (World Scientific, 2005), p. 5.
2. C. N. Yang, Modern physics and warm friendship, in *Lattice Dynamics and Semiconductor Physics*, eds. J. B. Xia *et al* (World Scientific, 1990).
3. C. N. Yang, in *Proc. Int. Conf. "Enrico Fermi and the Universe of Physics*, Rome, September 29–October 2, 2001, eds. C. Bernardini *et al* (ENEA – Ente per le Nuove tecnologie, L'Energia e l'Ambient, Roma, 2003).
4. C. N. Yang, *Selected Papers 1945–1980 with Commentary* (World Scientific, 2005), p. 6.
5. J. Cronin, C. N. Yang and CP violation, in *Chen Ning Yang, A Great Physicist of the Twentieth Century*, eds. C. S. Liu and S. T. Yau (International Press, 1995).
6. 杨振宁,《读书教学四十年》(三联书店(香港)有限公司, 1985), p. 120.
7. C. N. Yang, *Selected Papers 1945–1980 with Commentary* (World Scientific, 2005), pp. 11–12.
8. H. Poincaré, *Science and Method* (Dover, 1952), p. 56.
9. C. N. Yang, *Selected Papers 1945–1980 with Commentary* (World Scientific, 2005), p. 20.
10. C. N. Yang, *Selected Papers 1945–1980 with Commentary* (World Scientific, 2005), p. 67.
11. 杨振宁,《曙光集》(三联书店, 2008), pp. 364–365.
12. C. N. Yang, *Selected Papers 1945–1980 with Commentary* (World Scientific, 2005), p. 241.
13. 江才健著,《吴健雄》(复旦大学出版社, 1997), p. 193.

Reflections on Science & Society

https://doi.org/10.1142/9789819826636_0005

THEMATIC MELODIES OF TWENTIETH CENTURY THEORETICAL PHYSICS: QUANTIZATION, SYMMETRY AND PHASE FACTOR*

CHEN NING YANG
Chinese University, Hong Kong
and
Tsinghua University, Beijing 100084, China

Received 2 July 2003

It has been said that the 20th century was the century of physics. There are ample reasons to support this statement: It was in that century that man discovered, for the first time since our ancestors discovered fire, the second and the vastly stronger source of energy: nuclear power. It was in that century that man learned to manipulate electrons to create the transistor and the modern computer, transforming thereby human productivity and human lives. It was in that century that man learned how to probe into structures of atomic dimensions, discovering thereby the double-helix, a key to the secrets of life. It was in that century that man ceased to be earth-bound, taking first steps on the moon. In short, it was a century in which man made unprecedented progress on many fronts of human activities. And these progresses were largely ushered in by breathtaking advances in the science of physics.

It is hard not to be impressed by the decisive roles that the climatic developments in 20th century physics had played in human history. But decisive as they are in human history, they do not represent, in fact, the true glory of the development of the science of physics in the 20th century.

The true glory of physics in the 20th century lay instead in the deepened understandings of important primordial concepts which date from the beginning of human civilization: that of *space*, of *time*, of *motion*, of *energy*, and of *force*. In all of these primordial concepts, there have been profound revolutions in our understanding, revolutions that had brought forth a more beautiful, more subtle, more precise and more unified description of nature.

*Talk at TH-2002, International Conference on Theoretical Physics, Paris, 2002.

Reproduced from International Journal of Modern Physics A, Vol. 18, No. 19, pp. 3263–3272 (2003), DOI: 10.1142/S0217751X03016070

There have been in recent years studies of many aspects of the detailed history of twentieth century physics. It is not my purpose to delve into these subjects here. What I propose to do is instead to look into this history for the broad *motifs* of the developments, and to trace the three main strands that had persistently woven through its conceptual advances, appearing again and again in a variety of forms, singly or intertwined, like the *thematic melodies* of symphonic music. We shall see that these three melodies together define the tone and the flavour of the main developments of physics in the 20th century.

1. Quantization

The century opened with a paper by Planck (1858–1947) in which a constant, now called Planck's constant, designating the elementary quantum of action, was introduced. Like the opening bars in a prelude to symphonic music, it was to become the first thematic melody of the physics of the 20th century.

Planck was a very careful physicist. It was unusual for him to take the dramatically bold step of proposing a quantum of action. He did. But after more mature thinking about the proposal, he evidently got cold feet and started to backtrack in ensuing years. The torch then passed onto the younger generation: Einstein (1879–1955) first made his proposal about the photoelectric effect, and Bohr (1885–1962) followed by formulating his theory of quantized hydrogen atom. By 1918 when Planck gave his Nobel speech entitled *The Genesis and Present State of Development of the Quantum Theory*, he said

> *... If the various experiments and experiences ... from the different fields of physics provide impressive proof in favour of the existence of the quantum of action, the quantum hypothesis has, nevertheless, its greater support from the establishment and development of the atom theory by Niels Bohr ...*

He then outlined the triumphs and enormous frustrations of generalizing Bohr's theory, but ended his speech on a positive note:

> *... that which appears today so unsatisfactory will in fact eventually, seen from a higher vantage point, be distinguished by its special harmony and simplicity ...*

The eventual "higher vantage point" did arrive, through the development of quantum mechanics, but only after an agonizing period of great turmoil from 1913 to 1927. Pais, the distinguished historian of physics, has described this period as, borrowing from Dickens,

> *It was the spring of hope,*
> *it was the winter of despair.*

Perhaps the best characterization of the feelings of physicists during this period was Oppenheimer's (1904–1967) in his 1953 Reith lecture:

"Our understanding of atomic physics, of what we call the quantum theory of atomic systems, had its origins at the turn of the century and its great synthesis and resolutions in the nineteen-twenties. It was a heroic time. It was not the doing of any one man; it involved the collaboration of scores of scientists from many different lands, though from first to last the deeply creative and subtle and critical spirit of Niels Bohr guided, restrained, deepened, and finally transmuted the enterprise. It was a period of patient work in the laboratory, of crucial experiments and daring action, of many false starts and many untenable conjectures. It was a time of earnest correspondence and hurried conferences, of debate, criticism, and brilliant mathematical improvisation."

"For those who participated, it was a time of creation; there was terror as well as exaltation in their new insight. It will probably not be recorded very completely as history. As history, its recreation would call for an art as high as the story of Oedipus or the story of Cromwell, yet in a realm of action so remote from our common experience that it is unlikely to be known to any poet or any historian."

To illustrate the great up and down swings deeply felt by physicists during that period, let us look at a letter from Pauli (1900–1958) to Kronig (1904–1995) dated May 21, 1925,

Physics is once again at a dead end at this time. For me, at any rate. It is much too difficult.

And another letter written by Pauli to Kronig five months later:

Heisenberg's mechanics has restored my zest for life.

What happened that had excited Pauli in between the two letters was some vague perceptions by Heisenberg (1901–1976) through a historic groping "in the fog." In a remarkable passage in his old age, Heisenberg likened this groping in 1925 to mountain climbing:

you sometimes ... want to climb some peak but there is fog everywhere ... you have your map or some other indication where you probably have to go and still you are completely lost in the fog. Then ... all of a sudden you see, quite vaguely in the fog, just a few minute things from which you say, "Oh, this is the rock I want." In the very moment that you have seen that, then the whole picture changes completely, because although you still don't know whether you will make the rock, nevertheless for a moment you say, "... Now I know where I am; I have to go closer to that and then I will certainly find the way to go ..."

The vague perceptions of Heisenberg that day in 1925 was to lead to quantum mechanics, which was undoubtedly one of the greatest intellectual revolutions in human history.

2. Symmetry

A second thematic melody, symmetry, had its earliest origin in the 1905 paper of Einstein which introduced the theory of special relativity. This paper launched a great revolution that changed physicists' conceptions of space and *time* in a profound way. That this revolution was mathematically describable, in a very elegant way, by the four-dimensional symmetry between space and time was not known until the 1908 paper of Minkowski (1864–1909).

While Einstein himself was not originally impressed with the "superfluous learnedness" of Minkowski's work, he soon changed his mind and began to go further: He tried to greatly generalize the symmetry of special relativity, which in Minkowski's paper was mathematically formulated as the invariance of physical laws under Lorentz transformations. In his *Autobiographical Notes* written many years later, Einstein described how he had started on the idea of generalizing the invariance (i.e. symmetry):

> ... *that the basic demand of the special theory of relativity (invariance of the laws under Lorentz-transformations) is too narrow, i.e. that an invariance of the laws must be postulated also relative to non-linear transformations of the coordinates in the four-dimensional continuum. This happened in* 1908.

This larger invariance, after eight years of struggle by Einstein, finally gave rise to the theory of general relativity.

From today's vantage point, this Einstein–Minkowski–Einstein development was the first example of the theme of *symmetry dictates interactions* which we shall come back to later. At the time, this theme was not immediately picked up and further developed. It was to lie dormant for many years, while a variation of the theme was played out with the advent of quantum mechanics.

To follow this story we have to go back to melody one: quantization. With the great experimental developments of spectroscopy, and with Bohr's theory of 1913, quantum numbers entered into the very language of atomic physics. They were used to describe the states of a dynamic system. After the formulation of quantum mechanics it soon was realized that some of these quantum numbers are deeply related to the symmetry of dynamic systems, and that a beautiful and well developed branch of mathematics called group theory is the right approach to the concept of symmetry: E.g. the experimentally observed quantum numbers occur entirely naturally in group theory.

This new entry of symmetry into fundamental physics was, however, at first resisted by most physicists. The term "group pest" was coined to label this invasion by unfamiliar sophisticated mathematical concepts. It took half a dozen years for this resistance to dissolve, and with the developments of nuclear physics starting in the 1930s, and the later development of elementary particle physics starting in the 1950s, symmetry and group theory gradually became a dominant theme in physics: E.g. throughout much of the 1950s and 1960s, the efforts in elementary particle physics were directed mainly at finding the quantum numbers of the new particles and the symmetry behind them. A paragraph from a 1957 speech by the present author can serve to illustrate how symmetry was regarded in those years:

> *It was, however, not until the development of quantum mechanics that the use of the symmetry principles began to permeate into the very language of physics. The quantum numbers that designate the states of a system are often identical with those that represent the symmetries of the system. It indeed is scarcely possible to overemphasize the role played by the symmetry principles in quantum mechanics. To quote two examples: The general structure of the periodic table is essentially a direct consequence of the isotropy of Coulomb's law. The existence of the antiparticles — namely the positron, the antiproton and the antineutron, were theoretically anticipated as consequences of the symmetry of physical laws with respect to Lorentz transformations. In both cases nature seems to take advantage of the simple mathematical representations of the symmetry laws. When one pauses to consider the elegance and the beautiful perfection of the mathematical reasoning involved and contrast it with the complex and far reaching physical consequences, a deep sense of respect for the power of the symmetry laws never fails to develop.*

3. Phase Factor

Dirac (1902–1984) was one of the architects of quantum mechanics. He had erected the foundation of the mathematics of quantum mechanics on the noncommutativity of the dynamical variables p (momentum) and q (coordinate). So it was remarkable that in 1972 when he was 70 years old he said:

> *So if one asks what is the main feature of quantum mechanics, I feel inclined now to say that it is not non-commutative algebra. It is the existence of probability amplitudes which underlie all atomic processes. Now a probability amplitude is related to experiment but only partially. The square of its modulus is something that we can observe. That is the probability which the experimental people get. But besides that there is a phase, a number of modulus unity which can modify without affecting the square of the modulus. And this phase is all important because it is the source of all interference phenomena but its physical significance is obscure. So the real genius of Heisenberg and Schrödinger, you might say, was to discover the existence of probability amplitudes containing this phase quantity which is very well hidden in nature and it is because it was so well hidden that people hadn't thought of quantum mechanics much earlier.*

Phase was of course a concept well known in the theory of various kinds of waves in earlier centuries. Its entry into fundamental physics, however, had a tortuous history which we shall outline as follows:

(a) In 1918 Weyl (1885–1955) tried to geometrize electromagnetism through the introduction of a linear differential form $d\varphi$, after Einstein's geometrization of gravity through the introduction of a quadratic differential form ds^2. He identified $d\varphi$ with $A_\mu \, dx^\mu$ times a constant, and considered a "stretch factor"

$$\exp\left[-\frac{e}{\gamma}\int A_\mu \, dx^\mu\right], \tag{1}$$

which he applied to a particle transported along a four-dimensional path, in analogy with the parallel displacement of a vector in general relativity. Weyl's ideas were criticized by Einstein who pointed out that if a meter stick is continually stretched along its world line, there would be no standardization of meter sticks possible, a devastating criticism. Weyl was not able to answer this criticism.

(b) In 1922 Schrödinger (1887–1961) in a paper with the title *On a Remarkable Property of the Quantum-Orbits of a Single Electron* observed that the stretch factor of Weyl, when evaluated along a Bohr orbit, gives an exponent which is an integral multiple of a constant. He called this a remarkable property, saying

> *It is difficult to believe that this result is merely an accidental mathematical consequence of the quantum conditions, and has no deeper physical meaning.*

He then speculated on what value the constant γ should take. What is truly remarkable, in retrospect, was the fact that one of the possibilities Schrödinger mentioned was

$$\gamma = -i\hbar, \tag{2}$$

> *in which case the stretch factor would be equal to unity. ... I do not dare to judge whether this would make sense in the context of Weyl geometry.*

All of this in 1922! Had Schrödinger pursued this possibility, he might have discovered wave mechanics in 1922! In the event, he apparently abandoned this whole line of thinking, only coming back to it after he read, in late 1925, de Broglie's suggestion of wavelengths for particles. In a letter dated November 3, 1925, he wrote to Einstein as follows:

> *The de Broglie interpretation of the quantum rules seems to me to be related in some ways to my note in the Zs. F. Phys.* **12**, 13 (1922), *where a remarkable property of the Weyl 'gauge factor'* exp $\left[-\int \phi \, dx\right]$ *along each quasi-period is shown. The mathematical situation is, as far*

as I can see, the same, only from me much more formal, less elegant and not really shown generally. Naturally de Broglie's consideration in the framework of his large theory is altogether of far greater value than my single statement, which I did not know what to make of at first.

Two and a half months later, in early 1926, he submitted for publication his monumental paper creating wave mechanics.

Substituting (2) into (1) converts Weyl's stretch factor into a phase factor, and that is, in retrospect, the earliest entry of the thematic melody of phase into quantum mechanics, but only as a distant hesitant background refrain.

(c) After 1926 Schrödinger did not return to this theme to further develop it. In fact, he strongly resisted to introduce $i = \sqrt{-1}$ into his wave equations. It was Fock and London who recognized the necessity of i in quantum mechanical electromagnetism. In particular, in a paper in 1927 by London with the title, *Quantum-Mechanical Interpretation of Weyl's Theory*, which quoted the Schrödinger 1922 paper, the stretch factor, which has become the phase factor, was the centre of the discussion.

(d) Weyl in 1929 came back with an important paper that really launched what was called, and is still called, gauge theory of electromagnetism, a misnomer. (It should have been called phase theory of electromagnetism.) This amazing paper also discussed the 2 component neutrino theory which later in the 1950s became very important.

(e) Weyl's papers were rambling and philosophical. Few physicists of my generation read them. We all learned about Weyl's gauge theory of electromagnetism from the review papers of Pauli in the *Handbuch der Physik* (1933) and *Rev. Mod. Phys.* (1941). Pauli, however, did not emphasize Weyl's concept of a stretch factor metamorphed into a phase factor for electromagnetism. Thus the importance of the phase factor had to lie dormant for several decades.

4. Development

All three thematic melodies were introduced in the first decades of the 20th century. Together they played roles in the rest of the century very much like the roles of thematic melodies in music: through developments, variations and entwining.

In the late 1940s, many new particles were discovered in cosmic ray experiments. They were unexpected, so they were called "strange particles." How these strange particles interact between themselves and with known particles became naturally a topic of discussion. For a few years these discussion followed the idea of writing down Lorentz-invariant couplings, such as scalar meson with vector couplings, vector meson with tensor couplings, etc., and deriving observable consequences of each possibility. Such ideas lacked a principle, a *general principle of interactions.* In 1954 Yang and Mills tried to formulate such a principle through generalizing Weyl's gauge theory of electromagnetism. They observed that electric charge conservation is related through Weyl's theory to an invariance: gauge invariance, or gauge symmetry. So perhaps another conservation law widely discussed at that time, conservation of isotopic spin, should be related to a generalized gauge invariance, or gauge symmetry. In their own words of 1954:

> *The conservation of isotopic spin points to the existence of a fundamental invariance law similar to the conservation of electric charge. In the latter case, the electric charge serves as a source of electromagnetic field; an important concept in this case is gauge invariance which is closely connected with* (1) *the equation of motion of the electromagnetic field*, (2) *the existence of a current density*, *and* (3) *the possible interactions between a charged field and the electromagnetic field. We have tried to generalize this concept of gauge invariance to apply to isotopic spin conservation. It turns out that a very natural generalization is possible.*

This natural generalization was later called Yang–Mills theory, or non-Abelian gauge theory.

Another way that also led to the same generalization was summarized in the beginning sentence of the abstract to another 1954 paper by these authors:

> *It is pointed out that the usual principle of invariance under isotopic spin rotation is not consistent with the concept of localized fields.*

This sentence is overstated. It should be amended to read

> *It is pointed out that the usual principle of invariance under isotopic spin rotation is not consistent with the spirit of the concept of localized fields.*

Both ways of approaching the generalization emphasized invariance, i.e. symmetry. The phase factor aspect was of course there, (the whole development being a quantum mechanical theory), but was not at that time central to the generalization. In particular, the stretch factor (changed into a phase factor), which was central to Weyl's thinking, was not correspondingly generalized. It was only in 1974 when it was finally generalized, and was called a nonintegrable non-Abelian phase factor.

With the nonintegrable phase factor the entwining of two of the thematic melodies, symmetry and phase factor, became intimate and intrinsic. Reviewing this history it is amazing that both melodies were close to the heart of Weyl throughout his life. Unfortunately, when he died in 1955, he did not have the opportunity to witness the great fruition of his ideas. He had this to say about himself in 1949:

> *A lone wolf in Zürich, Hermann Weyl, also busied himself in this field; unfortunately he was all too prone to mix up his mathematics with physical and philosophical speculations.*

Today, we would say that his speculations were amazingly insightful, and had served to change the course of the history of physics.

Non-Abelian gauge theory did not impress the physics community in the 1950s. The theory was patterned after electromagnetism in which the vector boson, i.e. the photon, has zero mass. It was therefore generally believed that in non-Abelian gauge theory the vector boson would also have zero mass. Since there were no such mesons found, the theory was not convincing. In 1954, Yang and Mills discussed this question in the last paragraph of their paper, and concluded

> *In electrodynamics, by the requirement of electric charge conservation, it is argued that the mass of the photon vanishes. Corresponding arguments in the* **b** *field case do not exist even though the conservation of isotopic spin still holds. We have therefore not been able to conclude anything about the mass of the* **b** *quantum.*

Then in the 1960s an important new idea about symmetry called *symmetry breaking* was formulated by several authors. The idea allows for complete symmetry in mathematical formalism, but at the same time allows for breaking of the symmetry in physical manifestations. The possibility then exists for the vector boson in a non-Abelian gauge theory to have nonzero mass. This idea was vigorously pursued both theoretically and experimentally, starting in the early 1970s, resulting in what is now called the standard model, which has been spectacularly successful in describing the multitude of experimental results in elementary particle physics.

In non-Abelian gauge theory, the interaction is determined by symmetry (i.e. invariance with respect to gauge transformations). Hence the principle: *symmetry dictates interaction*. As such it picked up the Einstein–Minkowski–Einstein melody that had earlier determined gravitational interaction through the requirement of invariance (i.e. symmetry) with respect to coordinate transformations.

Although the standard model has been spectacularly successful, it has a big defect: the way symmetry breaking is applied is *ad hoc*, and not unique. Few physicists believe therefore that the standard model is the final theory. This problem, together with the lack of success of putting general relativity and quantum theory together, remain basic problems of fundamental physics at the close of the 20th century.

The three melodies, quantization, symmetry and phase factor are beautifully and subtly entwined in the path integral formalism of Feynman, modified by requiring the integrand to be a time-ordered product,

$$\text{propagator} = \int \exp\left[\frac{i}{\hbar}\,(\text{action})\right] d\,(\text{path}) \tag{3}$$

for non-Abelian gauge theory: Quantization is an essential strand through the presence of the quantum of action, $\hbar$. Phase factor is an essential strand through the presence of $i = \sqrt{-1}$, and symmetry is an essential strand because the action for a non-Abelian gauge theory has a subtle multiplicative property under a gauge transformation.

It is amazing that all three thematic melodies were evolved from primordial concepts in the cognitive history of mankind: quantization through the recognition of units of measurement; symmetry through the recognition of the beauty of geometrical forms; phase through the observation of the phases of the moon. Twentieth century physics assigned precise mathematical meanings to each of these concepts and put them together in the path integral form (3), expressing, we might say, much that is the best in the tradition of physics since Newton and Huygens.

https://doi.org/10.1142/9789819826636_0006

The Future of Physics — Revisited*

C. N. Yang
Tsinghua University, China

Published 21 July 2015

I

In April 1961 there was a big Centennial Celebration at MIT. That was when science and technology were making unprecedented progress for mankind and when the United States had just inaugurated her young and ambitious new President. Naturally it was a proud, joyous, even intoxicating Celebration. At the week-long event there was a panel discussion on "The Future of Physics", chaired by Francis Low, with four speakers in the following order: Cockcroft, Peierls, Yang and Feynman. It was originally understood that the talks were to be published by MIT, but somehow that never happened. Much later the talks by Cockcroft and Peierls were summarized by Schweber in his 2008 book "Einstein and Oppenheimer". My talk and Feynman's were published, respectively in 1983[1] and in 2005.[2]

In my talk I said,

> *"But since there seems to be too ready a tendency to have boundless faith in a "future fundamental theory", I shall sound some pessimistic notes. And in this Centennial celebration, in an atmosphere charged with excitement, with pride for past achievements and an expansive outlook for the future, it is perhaps not entirely inappropriate to interject these somewhat discordant notes."*

I then argued that to reach the present level of understanding of field theory, according to Wigner's counting, one must penetrate four levels of physical concepts formed out of experiments. Furthermore to reach deeper levels of penetration will become more and more difficult.

*Based on talk given at the Conference on 60 Years of Yang–Mills Gauge Field Theories, 25–28 May 2015, NTU, Singapore.
Editorial comment: We think this is a very interesting article. For the convenience of the readers, we reprint below the talks by Yang and by Feynman at that 1961 panel (App. A and App. B).

Reproduced from International Journal of Modern Physics A, Vol. 30, No. 21, 1530049 (2015), DOI: 10.1142/S0217751X15300495

> *"Here physicists are handicapped by the fact that physical theories have their justification in reality. Unlike the mathematicians, or the artists, physicists cannot create new concepts and construct new theories by free imagination."*

I was followed by Feynman, who began as follows:

> *"As I listened to Professor Cockcroft, Professor Peierls and Professor Yang, I found that I agreed with almost everything they said.* (But) *I don't agree with Professor Yang's idea that the thing is getting too difficult for us. I still have courage. I think that it always looked difficult at any stage."*
>
> *"One possibility is that a final solution will be obtained. I disagree with Professor Yang that it's self-evident that this is impossible."*
>
> *"What I mean by a final solution is that a set of fundamental laws will be found, such that each new experiment only results in checking laws already known."*

Feynman was one great intuitive theoretical physicist of my generation. Reading these passages today, I wonder

(1) what type of "final solution" he had in mind in 1961, and
(2) whether he still held such very optimistic views later in his life.

II

What have we learned in the fifty odd years since that 1961 panel?
A lot. Through very intensive collaborative efforts between theorists and experimentalists important conceptual developments were proposed and verified:

- One specific model of Symmetry Breaking
- Electroweak Theory
- Renormalizibility of non-Abelian Gauge Theory
- Asymptotic Freedom and QCD

Capping these developments was the dramatic experimental discovery in 2012 of the Higgs particle.

We have now a workable

$$\textit{standard model}, \text{ a } \mathbf{SU(3)} \times \mathbf{SU(2)} \times \mathbf{U(1)} \textbf{ gauge theory}.$$

Thus in these fifty odd years since 1961, we have reached *one more layer* of physical concepts, which is based on *all* the previous layers *plus* a large number of very large experiments.

III

Are there additional layers of physical concepts deeper than those that we have reached so far? *I believe yes, many more.*

When can we expect to reach the next level? *I believe in the distance future, if ever.*

Why are you so pessimistic? *I am not pessimistic, I am just realistic.*

References

1. C. N. Yang, *Selected Papers* (Freeman, 1983), p. 319; E. P. Wigner, *Proc. Amer. Phil. Soc.* **94**, 422 (1950) [Reprinted below as Appendix B].
2. M. Feynman (ed.), *Perfectly Reasonable Deviations from the Beaten Track: The Letters of Richard P. Feynman* (Basic Books, 2005), Appendix III [Reprinted below as Appendix A].

Appendix A

The Future of Physics‡

Richard P. Feynman

As I listened to Professor Cockcroft, Professor Peierls and Professor Yang, I found that l agreed with almost everything they said. I don't agree with Professor Yang's idea that the thing is getting too difficult for us. I still have courage. I think that it always looked difficult at any stage. On the other hand, I agree, as you will see, with something about this pessimism, and I don't think that I can add anything sensible to anything that the other speakers said. So in order to proceed, I have to add something that is not sensible, and if you will excuse me therefore, I am going to try to say something quite different than what they said.

First of all, to make the subject not infinite, I am going to limit myself very much and discuss only the problem of the discovery of the fundamental laws in physics — the very front line. If I were talking about that which is behind the front line, things like solid-state physics, and other applications of physics and so on, I would say very different things. So please appreciate this limitation of the discussion.

I do not think that you can read history without wondering what is the future of your own field, in a wider sense. I do not think you can predict the future of physics alone with the context of the political and social world in which it lies. If you want to predict, as Professor Peierls does, the physics a quarter of a century in the future, you must remember that you are trying to predict the physics of 1984.

‡As published in The Technology Review, 1961–1962. Extracted from "Perfectly Reasonable Derivations from the Beaten Track: The Letters of Feynman" (Basic Books, 2005), Appendix III.

The other speakers seem to want to be safe in their predictions, to they predict for 10, perhaps 25, years ahead. They are not so safe because you will catch up with them and see that they were wrong. So, I am going to be really safe by predicting 1,000 years ahead.

What we must do according to the method of prediction used by the other speakers is to look at the physics of 961 and compare it to the present, 1961. We must compare the physics even a century before the age when Omar Khayyam could come out the same door as in he went, to the physics of today as we open one door after the other and see rooms with treasures in them, and in the backs of the rooms five or six doors obviously going into other rooms with even greater treasures. This is a heroic age. This is an exciting time of very vital development in the fundamental physics and the study of the fundamental laws. It is not fair to compare it to 961, but to find another heroic age in the development of physics, the age of, perhaps, Archimedes, of Aristarchus; say, the third century B.C. Add 1,000 years and you find the future of their physics: the physics of the year 750! The future of physics depends on the circumstances of the rest of the world and is not merely a question of extrapolation of the present rate of progress into the future. If I go a thousand years, I have a difficult problem. Is it possible that it's all going to be over?

One of the most likely things, from a political and social point of view, is that we have soon a terrific war and a collapse. What will happen to physics after such a collapse? Will it recover? I would like to suggest that the physics, fundamental physics, may possibly not recover, and to give some arguments as to why not.

In the first place it is very likely that if there were sufficient destruction in the Northern Hemisphere the high-energy machines, which seem to be necessary for further research, would become inoperative. The machines themselves may be destroyed, the electrical power to operate them may be unavailable, and the industrial technology needed to repair or maintain them may no longer exist, at least for a while. Experimental physics techniques are the quintessence of our technological and industrial abilities, and so they must suffer some temporary setback.

Can physics slide back temporarily and then recover? I don't think so. Because, in order to have this heroic age an exciting one, one must have a series of successes. If you look at the grand ages of different civilizations, you see that people have an enormous confidence in success, that they have some new thing that is different, and that they are developing it by themselves. If one were to slide back, you would find for a while, no great successes. You would be doing experiments that were done before. You would be working on theories that "the ancients" knew very well. What could result would be a lot of mouthing and philosophizing; a great effort to do the physics in the sense that one should do it to be civilized again, but not really to do it. To write, instead, commentaries, that disease of the intellect, which appears in so many fields. Physics is technically too hard to recover immediately. There would be practical problems at that time that would occupy the attention of intelligent people. The difficulty is that there would be no fun in it. The new discoveries wouldn't come for a while. The other feature is that it would not be useful. No one has yet thought of a use of the results of the experiences we have with the high-energy particles. And finally, it is possible that antagonism is produced by the terrible calamity; there might be a universal antagonism toward physics and physicists as a result of the destruction which people might blame on the scientists who made it possible. Another thing to remember is that the spirit of research may not build itself up again because this spirit is concentrated in the Northern and advanced industrial countries, and this spirit does not exist fully in the other countries.

Well, I said 1,000 years. Maybe in that much time there will be another renaissance. What kind of machinery could there be for a recovery from this thing? (I said I wouldn't talk about anything sensible. I can't.) Some success somewhere must be the cause of a new renaissance. Where will this success lie? Perhaps in other fields. Perhaps in some other field than physics one would find a new age developing a success above "the ancients" and then getting a new confidence and growing. When this grows, it can pass its enthusiasm to physics. Or perhaps there would be a new aspect to physics, some other point of view, or some other completely different thing. That I cannot tell.

Another interesting possibility is that the renaissance may lie in some nation or people discovering a success by using a scientific attitude as a kind of morality, in society, government and business. You know what I mean — when someone says something, looking for what it is that they are saying, not why they are saying it. Propaganda would be a dishonest thing, and no one would pay attention to someone who would say something not for the content of the idea they want to get across, but because they want to show that they are big, or good, or some such reason. It is possible that if any success results from using such a scientific attitude, a country would be encouraged to go on from this success in its society and develop a re-interest in the scientific problems.

Well, now let's take the opposite view. Suppose there is no collapse. How, I don't know, but suppose there is no collapse. Then what? Suppose we can imagine a society somewhat like our own continuing for a thousand years. (Ridiculous!) What would happen to fundamental physics, the fundamental problems, the study of the laws of physics?

One possibility is that a final solution will be obtained. I disagree with Professor Yang that it's self-evident that this is impossible. It has not been found yet, but if you were walking through a building to get from one side to the other, and you hadn't yet reached the door, you could always argue, "Look, we've been walking through this building, we haven't reached the door, therefore there is no door at the other end." So it seems to me that we are walking through a building and we do not know whether it is a long infinite building, or a finite building, so a possibility is that there is a final solution.

What I mean by a final solution is that a set of fundamental laws will be found, such that each new experiment only results in checking laws already known, and it gets relatively more and more boring as we find that time after time nothing new is discovered that disagrees with the fundamental principles already obtained. Of course, attention would then go to the second line about which I am not speaking. But the fundamental problem will have been solved.

I would say that one thing that would happen if such a final solution were found, would be the deterioration of a vigorous philosophy of science. It seems to me that the reason that we are so successful against the encroachment of professional philosophers and fools on the subject of knowledge, and the way of obtaining new knowledge, is that we are not completely successful in physics. We can always say to such people, "That was very clever of you to have explained why the world just has to be the way we have found it to be so far. But what is it going to look like to us tomorrow?" Since they are absolutely unable to make any predictions, we see that their philosophy does not have real understanding of the situation. But if the solution is all present, how many people are going to prove it had to be four dimensions, it just had to be this way because of such and such and so forth. And so the vigor of our philosophy, which is a vigor which conies from the fact that we are still struggling, I think that may fail.

What other possibilities are there? Suppose that the building we are walking through is infinite, as Professor Yang feels. Then there will be a continual exciting unfolding. We will rush through this house, one door after another, one treasure after another. A thousand years! Three unfoldings in sixty years is fifty unfoldings in a thousand years. Is the world going to have fifty exciting revolutions of our basic physical ideas? Is there that much treasure in fundamental physics? If there is, it will become somewhat boring. It will be boring to have to repeat it twenty times, this fact that things change always when you look deeper. I do not believe that it can last 1,000 years of active investigation. Well, if it doesn't stop (I mean if you can't get the final solution), and if I can't believe that it will keep on being excitingly developed for fifty revolutions, what else is there?

There is another possibility, and that is, that it will slow up. The questions will become more difficult, How will it look then? The strong couplings are analyzed, the weak couplings partially analyzed, but there are still weaker couplings that are harder to analyze. To obtain useful experimental information has become extremely difficult because the cross-sections are so tiny. Data comes in slower and slower. The discoveries are made more and more slowly, the questions get harder and harder. More and more people find it a relatively uninteresting subject. So it is left in an incomplete state with a few working very slowly at the edge on the question of what is this third-order tensor field that has a coupling constant 10 to the power of −30 times smaller than gravity?

It is possible, of course, that what we call physics will expand to include other things. I believe, for example, that physics will expand, just as Professor Peierls says, into the studies of astronomical history and cosmology. The laws of physics, as we presently know them, are of this kind. Given the present condition, what is the future? The laws are given by differential equations in time. But there must be another problem: what determines the present condition? That is, what is the whole history of the development of the universe? One way to see that this may someday be a part of physics and will not always be called astronomical history is to note there is at least a possibility that the laws of physics change with time. If the laws of physics change with absolute time, then there will be no way to separate the problems of formulating the laws and of finding the history I think that it is very likely that cosmological problems will be enmeshed in physics.

Finally, I must remind you that I limited myself to talking about the future in fundamental physics. I'd say that there will be an important return from the front line into the applications and the development of the consequences of the laws. This will be a very exciting thing, and I would say quite different things about its future than I would say about the future of the fundamental laws.

We live in a heroic, a unique and wonderful age of excitement. It's going to be looked at with great jealousy in the ages to come. How would it have been to live in the time when they were discovering the fundamental laws? You can't discover America twice, and we can be jealous of Columbus. You say, yes, but if not America, then there are other planets to explore. That is true. And if not fundamental physics, then there are other questions to investigate.

I would summarize by saying that I believe that fundamental physics has a finite lifetime. It has a while to go. At the present moment it is going with terrific excitement, and I do not want to retire from the field at all. I take advantage of the fact that I live in the right age, but I do not think it will go on for a thousand years.

Now, to finish, I would summarize two points. First, I did not talk about applied physics or other fields, about which I would give a considerably different talk. And second, in these modern times of high-speed change, what I am forecasting for a thousand years will probably occur in a hundred.

Thank you.

Appendix B

The Future of Physics[†]

C. N. Yang

In the last four or five years much effort and attention have been devoted by theoretical physicists to the analytic continuation from physically observable experience into unphysical regions. In particular, it has been tried by extrapolation to study properties of the singularities in the unobserved region. Such attempts have been beset from the beginning with great difficulties. But interest in them maintained. In a similar spirit, this morning we try to pursue a parallel approach: By extrapolation we want to look beyond our past experience and learn something about the so far

[†]Panel discussion by C. N. Yang at the MIT Centennial Celebration, April 8, 1961.

unobserved future development of physics. We cannot hope for any real success in this pursuit. But I believe we all agree the attempt is highly interesting.

By all standards the achievements in physics in the twentieth century so far have been spectacular. Whereas at the turn of the century the atomic aspect of matter was just emerging as a new subject of study, we have today progressed dimensionwise by a refinement of a factor of a million: from atomic dimensions to subnuclear dimensions. Energywise the progress is even more impressive: from a few electron volts to multibillion volts. And the power and ingenuities of the experimental techniques have fully kept pace with the increased depth of the physicists' inquiries. The influence and impact brought about by the advances in physics upon other sciences — upon chemistry, astronomy, and even upon biology — have been important beyond description. Similar influence upon technology, upon human affairs, has been so preoccupying in the postwar years as to need no further emphasis here.

But it is not in these influences that the glory of physics and the heart of the physicist lie. It is not even in the continued enlargement of the domain of physical experimentation, important as it is, that the physicists take the greatest pride and satisfaction. What makes physics so unique as an intellectual endeavor lies in the possibility of the formulation of concepts out of which, in the words of Einstein,[1] a "comprehensive workable system of theoretical physics" can be constructed. Such a system embodies universal elementary laws, "from which the cosmos can be built up by pure deduction."

Judged by such a lofty and rigorous standard, the sixty years of the twentieth century stand out as nothing short of heroic; for besides the numerous important discoveries that *widened* our knowledge of the physical world, this period has witnessed not one, not two, but three revolutionary changes in physical concepts: the special relativity, the general relativity and the quantum theory. Out of these conceptual revolutions were constructed *deepened*, comprehensive and unified systems of theoretical physics.

Endowed with such a distinguished heritage of the recent past, what are the prospects for its future?

Surely the rapid widening of knowledge will continue, both in what Professor Peierls refers to as the foundation of physics and in what he calls behind the front line.

In the former our present knowledge is sufficient to enable us to say with some certainty that great clarification will come in the field of weak interactions in the next few years. With luck on our side we might even hope to see some integration of the various manifestations of the weak interactions.

Beyond that we are on very uncertain grounds. To be sure we can already formulate a number of questions, the answers to which seem to us at this moment to be crucial: How should one treat a system with an infinite degree of freedom? Is the continuum concept of space time extrapolatable to regions of space 10^{-14} cm to 10^{-17} cm, and to regions smaller than 10^{-17} cm? What are the bases of the invariance under charge conjugation, and the invariance under isotopic spin rotation, both of which, unlike space–time symmetries, are known to be violated? What is the unifying basis of the strong, the electromagnetic and the weak interactions? What is the role of the gravitational field relative to all these? The list can go on, but we are not even sure that these questions are meaningful as we here phrased them: in fact much of the progress in physics had developed from the very recognition of the meaninglessness of some previously asked questions.

Of one thing, however, we can be sure. The accumulation of knowledge will proceed at a very rapid pace. We need only remind ourselves that not so very long ago, the time scale of discoveries in physics was measured in years, if not in decades. The Michelson–Morley experiment, for example, was first performed in 1881, then repeated with greater precision in 1887. To explain the negative result of the experiment Fitzgerald invented a contraction hypothesis in 1892, and then Lorentz invented the Lorentz transformation in 1902, culminating in Einstein's special theory of relativity in 1905. Imagine that Michelson's first experiment were done today!

The general awakening of mankind to the importance of science and the amazing ingenuity of the human mind at technological creativity virtually ensure for us further quickening of the pace of the experimental sciences.

But where shall we stand with respect to the "comprehensive system of theoretical physics" that was referred to a few minutes ago? Can we reasonably expect further success in the glorious tradition of the first sixty years of the twentieth century?

If it is difficult to locate singularities of functions by extrapolation, it is as difficult to predict revolutionary changes in physical concepts by forecasting. But since there seems to be too ready a tendency to have boundless faith in a "future fundamental theory," I shall sound some pessimistic notes. And in this Centennial celebration, in an atmosphere charged with excitement, with pride for past achievements and with an expansive outlook for the future, it is perhaps not entirely inappropriate to interject these somewhat discordant notes.

First, let us emphasize again that the mere accumulation of knowledge, while interesting, and beneficial to mankind, is nonetheless quite different from the aim of fundamental physics.

Second, the subject matter of subnuclear physics is already very remote from the direct sensory experience of mankind, and this remoteness is certainly going to increase as we delve into even smaller dimensions. A direct and graphic proof of this can be easily found in the growing physical size of the laboratories, the accelerators, the detectors and the computers.

What we call an experiment today consists of elaborate operations with elaborate equipments. To make any sense at all of the results of an experiment, concepts have to be formulated on all levels between our direct sensory experience and the actual experimental arrangement. The difficulty inherent in this state of affairs is as follows. Each level of concepts is connected with and in fact built upon the previous levels. When inadequacies manifest themselves, one must reach for greater depth by examining the whole complex of previous concepts. The difficulty of this task rapidly diverges with the depth of the considerations, much like in chess playing, it becomes increasingly difficult to practice always examining one more move ahead as one's skill improves.

According to Wigner's counting,[2] to reach our present investigations in field theory one must penetrate at least four levels of concepts. The details of the counting may be a subject of discussion, but there is no denying that what we envisage as a construction of a deeper and more comprehensive theoretical system represents at least one more level of penetration. Here physicists are handicapped by the fact that physical theories have their final justification in reality. Unlike the mathematicians, or the artists, physicists cannot create new concepts and construct new theories by free imagination.

Third, Eddington[3] has once given the example of a marine biologist who uses a net with 6 inch holes and who after careful and long studies formulates the law that all fish are larger than 6 inches. If this imaginary example sounds ridiculous, we can easily find examples in modern physics where because of the complexity and indirectness of the experiments, it has happened that one does not realize the selective nature of one's experiments, a selection that is based on concepts which may be inadequate.

Fourth, in the day-to-day work of a physicist it is very natural to implicitly believe that the power of the human intellect is limitless and the depth of natural phenomena finite. It is of course useful, or as is sometimes said, healthy, to have faith so as to have courage. But the belief that the depth of natural phenomena is finite is inconsistent and the faith that the power of the human intellect is limitless is false. Furthermore, important considerations have also to be given to the fact that psychological and social limitations on the development of the creative ability of each individual may be effectively even more stringent than natural limitations.

Having voiced these cautionary remarks, we must ask are they relevant to the development of physics, say, in the remaining forty years of this century? We cannot know the answer to this question now, but let us hope that it is in the negative.

References

1. A. Einstein, *Essays in Science* (Philosophical Library, New York, 1934).
2. E. P. Wigner, *Proc. Amer. Phil. Soc.* **94**, 422 (1950).
3. A. S. Eddington, *The Philosophy of Science* (MacMillan, New York, 1939).

https://doi.org/10.1142/9789819826636_0007

C. N. Yang on Teaching and Research in Physics

Yu Shi
Department of Physics, Fudan University, Shanghai 200433, China
yushi@fudan.edu.cn

David Waxman
Center for Computational Systems Biology, Fudan University, Shanghai 200433, China
davidwaxman@fudan.edu.cn

Published 28 April 2016

This document is based on five conversations between Prof. C. N. Yang and others in Beijing in 1986. In the conversations, Yang gave his views on the state and development of physics at that time, and the relationship between physics and philosophy. The conversations also contain Yang's reminiscences on the creation of Yang–Mills theory and his advice to young people, especially those in China.

In the early summer of 1986, Professor Chen Ning Yang was invited by the Graduate School (Beijing) of the University of Science and Technology of China to give a series of five lectures. China was just beginning to open up, and universities did not have adequate halls for this purpose. As a consequence, the lectures were given in the Friendship Hotel.

The audience consisted of faculty and graduate students from universities all over China. At the end of each lecture there were informal question (Q) and answer sessions, for which proceedings were later published in Chinese.[1]

Thirty years later, these question and answer sessions make for very interesting reading, from a historical perspective, and also from the perspective of the present. The following is our translation.

Reproduced from International Journal of Modern Physics A, Vol. 31, No. 12, 1630016 (2016), DOI: 10.1142/S0217751X16300167

Q: Please tell us about the goal of your lectures.

C. N. Yang: The subjects of the lectures were quite different, I think, from what was expected. It may have been thought that I would teach a lot about gauge field and high energy physics. If so, the audience was surprised. The contents of the lectures included neutron interference, the Aharonov–Bohm effect, flux quantization, holography, free electron lasers, quasi-crystals, high-energy elastic scattering, Dirac monopoles, fiber bundles, and non-Abelian gauge fields.

I intentionally did the lectures this way because of two goals I had for these lectures.

Firstly, I am interested in these subjects, many of which reflect the genuine spirit of 20th century physics. Some of these subjects are still under development, such as quasi-crystals and free electron lasers, and there will be great advances in the next twenty or so years and I thought that it would be worthwhile to introduce them to you.

Secondly and more importantly, through the selection of the topics, I hoped to inform the audience that they had better pursue new subjects in physics. What I wished to emphasize is distinct from the spirit of the teaching of physics that has taken place in China in the last few decades. Frankly speaking, physics students in China have been largely accustomed to sterile reading. I think the situation should be urgently changed. The whole social environment, arising from the attitudes of their parents and the media, has misled students, so that they have followed the wrong direction. Consequently, there are many students who are hard-working, well trained and very knowledgeable about a specific part of the subject, but their breadth of knowledge is narrow, and they have tended to develop expertise in highly mature aspects of the subject. This is very harmful. One of my goals, in the choice of the subjects I discuss here, is to introduce my viewpoint about physics to you. It is not necessary that everyone will be interested in all of the topics, however, if you think about what kind of topics these are, then hopefully you will begin to appreciate which physical problems are worth attention and further investigation.

In their studies, many students form an impression that physics is just calculation. Calculation is, of course, a part of physics, but not the most important part. The most important part is related to physical phenomena. Most of physics follows from observation of phenomena. Physical phenomena are the ultimate origin of physics. Without contact with phenomena, it is not impossible for a student or researcher to produce important work, but it is very probable that what they do is in the wrong direction and purely formal; they will miss the key point of physics. All the important physicists I know attach great importance to real physical phenomena. This is the strong impression I formed after I went to the USA to study. When I was a graduate student at the University of Chicago, I watched Fermi and Teller pay great attention to physical phenomena. Sometimes they performed a complicated calculation, but they were strongly focused on physical reality. During the period of 1948 to 1949, Fermi investigated renormalization, but he did not pay more attention to this than to physical phenomena. Renormalization was only one of many physical problems he considered. There is now a common situation that many people play games with the mathematical structures of physics, at the expense of studying real phenomena.

I have spoken about the teaching of physics in China many times in the last few years. I have given a speech "40 years of learning and teaching", which has been collected into a volume with the same title. I talked about it in Hong Kong in 1983, and later in Beijing and also Shanghai. It includes the following comments: "I have visited China many times since 1971, and discovered that there are the so-called 'Four Courses of Mechanics' (classical mechanics, electrodynamics, statistical mechanics and quantum mechanics) in physics departments in the universities, which depress the students. No one denies the importance of the 'Four Courses of Mechanics', which form the backbone of physics, but physics with only a backbone is a skeleton. Physics needs a backbone, but it also needs flesh and blood. Only with flesh and blood can physics come alive."

I think this topic touches on a very important problem. In the West, especially the USA, young people often lack training, but they have a spirit of fearing nothing, and are fond of thinking about new things, which are often close to experiments and real phenomena. I hope all of you pay more attention to new things, to living things and to things closely related to phenomena.

How could you actually achieve this? I have a concrete suggestion. There is a magazine entitled *Physics Today* in America, which is very good. On the first few pages of each issue, recent progress in physics is often introduced. These articles are well written and demonstrate deep thinking. The topics of recent progress that are introduced are, mostly, closely related to experiments, though some are purely theoretical. These articles are very readable and are only introductions, without going into details. For myself, I often read these articles to get to know about new progress of subjects I have heard of, or to get to know from which papers I can read about the new developments. There are many such articles with wonderful things mentioned. I suggest that graduate students and faculty members jointly look at new topics in each issue. In the beginning the study may not be very deep, only gaining knowledge about what is written. If, however, someone is interested in a particular topic, and is willing to read more of the literature on this, and present on this, then the whole group can go deeper. Many of the topics I have lectured on have been covered by *Physics Today* in the last 5 to 10 years. If, in a department of physics, there is a seminar series on the reports in the introduction of each issue of *Physics Today*, then this department will be doing a very good job at keeping up with recent developments.

Q: Please tell us about your expectation regarding the development of physics.

C. N. Yang: Different physicists have different viewpoints on this question. Let me tell you my personal ideas.

There has been great progress in physics in the 20th century. This was because of developments in experimental techniques. At the end of the last century, experimental techniques reached such a level that people were able to perform detailed spectroscopic investigations on molecules, atoms and nuclei. Spectroscopy has provided a large amount of data on atoms and molecules, and many problems that attracted attention in the first quarter of the 20th century came from spectroscopy. Indeed, this led to the formulation of quantum mechanics. In the 1930s, accelerators promoted the development of a new field, namely nuclear physics. Following the dropping of the atom bomb, governments of many countries linked progress of physics to the future of their countries, after the Second World War. As a consequence, a large amount of money was put into physics research. Together with industry, this created conditions for great progress in various aspects of physics. There have been astonishing developments in two important fields, namely high energy physics and condensed matter physics. However, further development of high energy physics is facing difficulty. It is difficult to do high energy experiments, as too much money is needed. This does not mean there will not be important achievements in high energy physics. For example, I have no doubt that if found, the discoveries of W and Z particles will become cornerstones of the subject. But the number of people working on high energy physics will decrease over time, and the achievements of each person in this field will also decrease. Nowadays, in conferences on high energy physics, it is common that not many important new results are reported. I suppose high energy physics, in the next 30 years, will be in a difficult period. This does not mean there will not be important work. Neither does it mean there will not be people working on it. Nevertheless the situation will not be a prosperous one. Although there are many smart people working on theoretical high energy physics, the theories will not be verifiable in experiments in the near future. Indeed the modern theoretical working style is very different from the past, and working with the door closed (on phenomena) has become inevitable.

On the other hand, condensed matter physics and areas related to technology will make great progress. It is relatively easy to have achievements in these areas.

Q: What is the current situation in theoretical physics? What period in the past is it comparable with?

C. N. Yang: Physics is very broad, and differences with the past lie in every direction. However, in most areas, the spirit is the same as in the past. For instance, in solid state physics, there have been great advances in experimental techniques, and there have been qualitative changes in methods of investigation. You cannot work on it at home, as was the case one hundred years ago, but its spirit is still very close to that of the subject 50 years ago.

Fundamental theoretical physics is based on particle physics. As I said above, particle physics experiments are increasingly expensive, and will inevitably decline in the next 30 years. With fewer and fewer experiments, the people remaining will be mostly theoretical. Some of them are very smart and it is inevitable that the field will become more and more mathematical. Already, fundamental theoretical physics is very mathematical. On the other hand, field theory and statistical mechanics are gradually being connected together. Statistical mechanics is closely related to condensed matter physics. Therefore, field theory is now increasingly connected with statistical mechanics and condensed matter physics. People at the starting stage of their career should pay attention to these points. As to the conclusion, it depends on what you like, what you have learned, and what opportunities you have.

Q: What problems do you think there are in scientific research in China? How can young people join research?

C. N. Yang: Let me focus on some general viewpoints that do not necessarily apply to everyone. In China, there are many able, creative and persistent people. No one doubts that China can be at the forefront of developments in physics. But there are also many problems in China. There are problems in industry, in city construction, in education, in living, in families, etc. There are many problems. Why, with so many smart and talented people, are there still so many problems? The answer is very simple and is unique. That is, China is poor. So earning money is the number one thing, at the levels of the individual and country. This does not mean everyone should go on to earn a lot of money. Instead, it means that the whole society should be economically successful. This is an absolute necessity. For many years I have regarded this as the most important thing for China. One should have it in mind when making important decisions. Hence I agree with a slogan in China: Twice double the total productivity of industry and agriculture by the year 2000. This slogan is rational, instead of irrationally making the goal too high. It will be remarkable to realize this goal in the year 2000. This will not only be important for China. It will also tell developing countries that are similar to China, that to make as great a development as China is possible if the goal is clear and achievable. By 2000 everything will be different in China. For the overall benefit of every individual, my understanding is that everyone should try to cooperate to achieve this goal. If an individual has made a contribution to the process of realizing this goal, they will recall it with pride in the future. For many years, I have told people studying physics not to enter high energy physics unless you feel you have to. High energy physics has nothing to do with China's twice doubling, and even has a negative contribution, as high energy physics is very costly. This does not mean high energy physics has no importance. Certainly it has importance, but the problem with China is that of "twice doubling" rather than that of high energy physics.

I feel that too many Chinese students in primary schools, middle schools, and universities, and even in graduate schools, overvalue studying. Studying is a tradition of China, and it has its advantages. But just because of this, it is not so easy to judge its true value. In the West, especially in the USA, studying is not regarded to be as important as it is in China. In China, especially in cities, parents wish their children to enter good middle schools, good universities, and obtain master degrees and doctoral degrees, and even higher degrees if there were any. This is harmful to China's economy as well as to the children themselves, as this is not achievable by every child. Here, you are all graduate students, hence you are all good at studying. If you are very happy in doing this then you may think about the problems I have lectured on. I do not agree with the very hard studying that occurred in the past in China. If you are very unhappy with study, you may think about whether doing something else is better for you and for society. For example, maybe you can make a better contribution if you go to a small factory, as you have some knowledge in physics, can speak a foreign language, and know something about the world. With such conditions, studying physics without joy may not be the best way forward.

Q: In "40 years of learning and teaching", you say the probability of success is higher if a young person enters a new field and grows together with it. What fields do you think have good prospects?

C. N. Yang: Quasi-crystals is an example that I have already mentioned. It is a new subject, and some of the basic concepts have not been made clear. Any new direction, with broad connections, has good prospects. If quasi-crystals could only be produced under very special conditions, the prospect for the field's development and impact would not be very good. But such is not the case of quasi-crystals, as we know.

Another example is if a new direction of experimentation occurs, as when two fast moving giant nuclei hit each other. This must lead to many new and interesting phenomena. Such phenomena may not be the most fundamental, but are still worth studying.

Yet another situation is the technology that is developed in a new field, which enables people to study phenomena that could not be previously studied. For example, now people can produce high powered lasers, with pulsed electric fields that are stronger than the electric fields inside atoms. Hitting an atom with such lasers can strip a whole shell of electrons. Such experiments are still very primitive. Entering such a field can easily lead to success.

If you work on renormalization of quantum electrodynamics, it is not easy to succeed, as it has a history that spans decades, and many very smart people have worked on it. How can you guarantee that you can do better than these people? This is not to say that working on such topics should be stopped. I am answering a question about which field it is easier to succeed in. To sum up, a field, new in either concepts or technology, or a new experimental direction, has better prospects than a well-trodden field.

Q: How should a physicist regard philosophy?

C. N. Yang: The word philosophy has many different meanings. In articles by Western physicists, there are two different meanings, one is the philosophy referred to by philosophers, another is a kind of view on physical problems, at a long or intermediate distance.

Q: Does the second philosophy refer to the epistemology?

C. N. Yang: No. Epistemology, as said by Shoichi Sakata, is a true philosophy, of the first kind. To give an example of the second kind, someone says: "Grand unified theories cannot possibly succeed." You say: "I agree." Then the first person says: "We have the same philosophy." No matter what words are used, the second kind of philosophy represents a viewpoint, namely, what kinds of problems receive attention. This has an important long-term influence. It decides what questions people like to raise, what questions they do not like to raise, what questions they would like answered, and what questions they do not care about. Also, the second kind of philosophy influences what methods people like to use, to solve a problem once the question arises. This sort of philosophy has much to do with one's style and preferences, and has a decisive impact on one's long term achievements in research. Everyone develops a philosophy of the second kind, out of their experience. Everyone should be reminded that it has important effects on their research, and they should be aware of it, to some extent.

As to the first kind of philosophy, I think it has a one-way relationship with physics. Physics influences philosophy, but philosophy never influences physics.

Q: Einstein thought he had been strongly influenced by Hume's and Mach's philosophy.

C. N. Yang: I do not agree with this saying. I think his success was because of the second kind of philosophy rather than this reason.

Studying physics is like looking at a large painting. The structure of the whole of nature is like this painting. There are several ways of looking at a painting. Sometimes one should combine different ways of looking. One way is to study it in detail, at a close distance. Since the painting has been made very carefully, different parts are different from each other, and you have to carefully study their details. Another way is to look at it from a great distance, where you may notice some patterns you cannot see at shorter distances. This is what is present at large scales. Certainly there are also views at intermediate distance. Physics needs views captured at far, intermediate and short distances. Certainly, if you can see a pattern that can be seen from a great distance, you make a big contribution. But this possibility is very small, almost impossibly so. You thus have to start at a short distance. Generally, the direction of knowledge is from near to intermediate to great distances, rather than the other way around.

An example of the above is that after its foundation, quantum mechanics strongly influenced philosophy, but neither Heisenberg nor Schrödinger started from philosophy. Instead, they started with atomic spectroscopy when they formulated quantum mechanics.

I entirely disagree with the idea of Sakata. I think Sakata has made some contributions to physics, but not out of his philosophy, instead, they originated in his understanding of physical reality. I disagree with his self-said origin in philosophy. His way of starting with philosophy went nowhere. I think he would have achieved more had he used less philosophy.

Q: What do you think of superstring theory?

C. N. Yang: Superstring is a topic of great current interest in theoretical high energy physics. I estimate that there are about 100 people with PhDs working on it. I hardly believe this theory will turn out to be right in the end. The basic concept in high energy theory is the field. This was started by Michael Faraday and through James Maxwell, and to the present, the idea of field was developed with many ups and downs, with countless experimental tests. The superstring started by generalizing the notion of field, without comparison with experiments. Now there are many papers on superstrings, but none of them has anything to do with experiments. Very likely it is a castle in the air, a dream. In Stony Brook, a graduate student asked me whether he should work on superstrings. My answer was the following. There are many important unsolved problems in high energy physics. There are many beautiful mathematical structures in superstring theory. If you are very interested in this subject and are very good at doing it, with excellent intuitive understanding of differential geometry and topology, you could work on it. But if you regard this direction to be certainly right, then you will be disappointed in the future, because the idea of superstrings has too little contact with real physics. Having little contact with real physics will not necessarily make the subject unsuccessful, but the chance of success is quite small. The mathematical structure of superstrings is very beautiful, and if you work on it, you will absorb these beautiful aspects. These may be helpful to your developing some methods to resolve some real questions. From this standpoint, you could go to do some work on superstrings. My words to the graduate student include my attitude to all of this class of questions concerning pure structures. It is very rare for an idea from abstract mathematics to become very successful in physics. You must be aware of this. Otherwise you could be very disappointed later on.

One can look at the connections of mathematics, theoretical physics and experiments in the following way:

4. Mathematics
3. Theoretical structures of physics.
2. Combining theories and experiments.
1. Experiments

Parts 2 and 3 together constitute theoretical physics. A pure theoretical structure is connected to experiments through the second part. A pure structure will lose its position in physics and disappear if it cannot be related to experiments.

The value in physics ultimately relates to experiments. Superstrings have not yet been related to experiments, as in the second part.

If you ask me whether I, myself, will work on superstrings, my answer is that I would never work on such a subject. If I can make myself understand it in two weeks, I will spend two weeks. But now it has become so complicated that I do not believe that I can work out results known to others in less than half a year. Because there are many smart people in this field, half a year is a very large investment, And it is not close to the physics I like. So I will not work on it. I know that I can do the kind of mathematical calculations required for superstrings. But if I were a graduate student, and my knowledge of physics had reached my current level, I would certainly go to pure mathematics. There are many beautiful things in pure mathematics. I should use my time and ability to work on the things that really advance mathematics, rather than go to work on things that neither advance physics nor have long-term value for mathematics.

Q: Now there are also people doing phenomenological work.

C. N. Yang: Those are far-fetched, far-fetched and far-fetched. Now that there is a pure structure, some people want to connect it to experiments in an ad-hoc way. As with superstrings, supersymmetry has not yet been connected with experiments.

Q: Please elaborate your attitude to supersymmetry.

C. N. Yang: Supersymmetry is very beautiful. Theories such as supersymmetry will continue to develop in the future, since the notion of symmetry is more and more important in physics. Since fermions and bosons are not yet treated symmetrically, contemporary theories are obviously imperfect.

The left-hand side of Einstein's gravity equation includes $R_{\mu\nu}$, while the right-hand side includes $T_{\mu\nu}$. Einstein thought the left-hand side is very beautiful, but the right-hand side is not good. He said that the left-hand side is made of gold, while the right-hand side is made of mud. He wanted to turn the matter contribution, on the right-hand side, also into something geometric, and move it to the left-hand side. To geometrize the matter contribution needs a unification of fermions and bosons.

The basic spirit of supersymmetry is very good, and has something very beautiful. The first time when I saw papers on supersymmetry, I did not believe it, as I thought the Feynman diagrams of fermions and bosons are different. In a certain theory, if the masses of fermions and bosons are equal in the lowest order approximation, then the masses will become unequal in higher orders. But my thinking was incorrect: there are some field theory models where the masses of fermions and bosons are equal in every order of perturbation. Hence it does have some beautiful aspects in some key points, but it still does not have anything to do with experiments, after more than 10 years.

If you ask another question, whether you should enter this field, I think you should be cautious. If you enter another newer field rather than this one, and if you are both interested and good at the subject, then it is easier for you to have good achievements. In a mature field, where many smart people have done a lot of work, what reason do you have for thinking that you will do better than them? This is like panning for gold, which is certainly better done in a new gold mine. This does not mean that nothing can be panned out of an old gold mine. But the possibility is lower. So I recommend panning in a new gold mine rather than an old one.

Q: What do you think of supergravity theory?

C. N. Yang: Einstein's gravity theory has a close relationship with experiment. Nevertheless he was not entirely satisfied with it. I have said that he wished to geometrize the matter contribution on the right-hand side, and move it to the left-hand side. It can be thought that supergravity is a beautiful proposal to resolve this problem. But it still has not been connected with experiments.

In Stony Brook, a graduate student from China wished to work on supergravity. I asked him: "are you very interested in supergravity?" He said: "yes." He learned geometry well. I said: "then you could work on it. You should obtain PhD in a shortest period of time, and pay attention to other subjects, such as solid state physics." He is now the best student of van Nieuwenhuizen. I believe he will excel in this field. I give this example to illustrate that if some people genuinely like to work on pure structures, and can do well, then let them to do it. But if you have not yet decided what to do, and are not as good as others at things like differential geometry, then I think you had better work on physics closely related to experimental phenomena, which is a safe way of training in theoretical physics. Supergravity is a pure structure, and will fade away if it cannot be related to experiments in 30 years.

Q: Where, in the four parts that you just talked about, does Yang–Mills theory belong?

C. N. Yang: There were students asking me about this in America. They said: "When you worked on gauge field in 1954, wasn't it also a pure structure having nothing to do with experiments?" Indeed it did not have much to do with experiments at that time, nevertheless, if you read our paper, you will find that it has close relations with two theoretical structures that have solid experimental foundations. These are conservation of isospin and the Maxwell equations. Now there is much work of adding flowers to silk, and for pure structures without experimental foundations, which become increasingly distant from experiments, it is very dangerous.

Ever since gauge field theory was put forward, I always paid attention to it, but did not write many papers. I always think it should not be carelessly changed to phenomenology. In the 1960s, there are people who thought it wonderful from the standpoint of pure structure, and did phenomenological work on it, amongst whom Sakurai was most notable. He thought that the ρ meson is a gauge particle. I did not agree. His way of doing things was very far-fetched. He wrote to me asking why

I was not interested in his work, in a complaining tone. Perhaps I did not reply to him. I was certainly interested in gauge fields, but thought his way of doing things was not good, so I had no way to answer him. Later, people introduced the idea of spontaneous symmetry breaking to gauge theories, thereby solving the mass problem of gauge particles without violating the spirit of symmetry. This is an important contribution. This experience tells us that many beautiful pure structures may be related to experiments, perhaps by a subtle revision. This is also the hope of people working on pure structures. However I warn you that the hope of success is very small.

Q: Are you satisfied with Higgs mechanism?

C. N. Yang: No. Everyone thinks this. It has its beauty, and fits current experiments. But nobody believes it is the final theory. Its idea is too ad hoc, and is without deep reasons of physics and mathematics, so it will be replaced by other theories. But it is very useful, for now.

Q: What do you think of the unification of gravitational fields and gauge fields?

C. N. Yang: Comparing their formulae, one can find that they are very similar. It is no problem that they have close relationship. But what kind of relationship is still a controversial question.

The $F_{\mu\nu}$ of gauge fields and the $R_{\mu\nu}$ of the gravitational field are both curvatures in geometry. The quantity $R_{\mu\nu}$ is the second derivative of $g_{\mu\nu}$, hence Einstein's gravitational field equation is a second order differential equation for the $g_{\mu\nu}$. Now that the equation of motion of a gauge field $\partial^{\mu} F_{\mu\nu} = \cdots$ is a first order differential equation, the gravitational equation should be a third order differential equation of $g_{\mu\nu}$. This is also an indication that Einstein's gravitational equation needs revision. In 1974, after the geometric structure of gauge fields was clear, I put forward a new equation for the gravitational field. But while I was able to write down the left-hand side of the equation, I was not able to write down the right-hand side. This problem has not yet been solved up to now.

Q: Please comment on the grand unified theory and the role of gauge fields in it.

C. N. Yang: Grand unified theory aims to extend the successful electroweak unified theory to also include the strong interaction. However, grand unified theory does not fit experiments. Nature is mysterious. Grand unified theory only simply extends a successful existing theory. Without a new idea, its lack of success is not surprising.

I think the direction of unifying more objects is correct, and I believe that most theoretical physicists agree to this. How they are to be unified in the future I don't know. No doubt gauge fields will play an important role in the unification, but perhaps it is not enough to only have this, perhaps there ought to be new ideas that have not been thought of.

Q: What comments would you like to make on sub-quark unified theory?

C. N. Yang: These are very speculative subjects. There are many papers, and they have been investigated for many years. It is not appropriate for young people, at the beginning of their graduate studies, to develop in such a direction. Rather, they should work on problems relevant to physical phenomena.

Q: What do you think of the fifth interaction?

C. N. Yang: Early this year some journalist reported a paper published in *Physical Review Letters*, hence the whole world knows there is the so-called fifth interaction. I do not believe such an interaction with the interacting range neither long nor short. If I were the referee, I would agree to publish this paper, but I don't think it would merit additional publicity.

You do not receive much news in China, so it is easy to be dominated by some fragments of the news. Physicists should have their own attitudes. Some attitudes may be wrong, and should be revised in time. But with their own attitudes, they would not be like weak trees that easily shake with the wind. Some people like to connect some strange structure with some rare experiments. It often happens, but the probability of success is very small.

Q: Please tell us what you think of the phenomenological theories of associated production.

C. N. Yang: Frankly speaking, work on this direction, abroad, is without any value, since the people working in this do not know what physics is. In China, in seeing a paper in *Physical Review*, one often immediately makes an effort to study it. This is often a disadvantage. Papers often come out in the following way. A writes a paper. Then B says it is wrong, and makes a revision. Then C says that B's paper is wrong, and makes a revision. You read the paper of C, and are trapped in others' puzzles without foundation. Regarding such things, you should study the original experiment. It is relatively easy to study new things. As it is new, it has not been made confusing by many papers. If you enter the field at this time, which is now close to experiments, then it is easy to obtain new interesting results. I have two suggestions: pay more attention to new subjects under development, and pay more attention to original phenomenon. If these two points are implemented, then it is easy to be connected with real physics.

Q: Please tell us what you think of chaos theory.

C. N. Yang: This is a new and interesting field. A few years ago, I suggested to some people in China that they work on it. Now it has had a history of 7 to 8 years. But it is still worth working on.

Q: Please comment on finite temperature field theory.

C. N. Yang: Finite temperature field theory is very interesting. This is a topic with depth. There are many papers on it, but I did not study them. For such a problem, my general attitude is the following. If I decide to work on it, I will start from the beginning, without reading others' papers first. Only if there are some difficulties, after working for some while, do I go to read others' papers. Only in such a way do I digest their work well. Around 1959, T. D. Lee and I wanted to discuss W intermediate bosons, and electromagnetic interactions of vector mesons. We worked from the start. After some time, we noted that many of the papers of others were wrong, though there were many. After a year, we became the top experts. Finite temperature field theory is a very good direction. If someone goes to work on it, I suggest that they work from the start, not necessarily reading others' papers, rather, reading them after working for some time. This is like arriving in a city for the first time. If you follow others from the start, you may still not know about the city after several journeys. If you try going around by yourself, the situation may be different.

Q: What do you think of the early Universe?

C. N. Yang: There have been many papers on it in the last 15 years, among which I suppose there are many interesting papers. However, I have not touched this field. Such questions are very speculative, which is not what I like working on. Some people like working on such a field, and are very successful. If someone is very suitable for doing such work, I think it is a good direction for them.

Q: My advisor asked me to work on superstrings, but I do not have the knowledge on supersymmetry and supergravity. It will need a long time to enter the field, time would be wasted if later it is found that the whole idea is problematic and I give up. On the other hand, I think life is determined by a biological field, and would like to research on it, but I fear that it could be castles in the air.

C. N. Yang: I understand everything you said except the biological field. If you have a very original idea, you may think about it in depth, but do not stick to it without limit, you should also pay attention to other things. Let me talk about my own experiences on this aspect.

When I was a graduate student in the University of Chicago, I thought about the following question. In relativity, there is a question of measuring a rigid body by using rulers, which was not yet clear after many years of debate. Now that only the observable subject can be established in physics (now I think this idea may not be always correct, and needs not to be regarded as too sacred), there are no rulers on the fundamental level, and there is no ordinary measurement, physics should not start from $ds^2 = g_{\mu\nu}\, dx^\mu\, dx^\nu$, as it is not fundamental. One should draw the worldline of every atom. These worldlines form a net in the four-dimensional space–time. Physics is the pattern of this net. At that time, I thought this idea is quite reasonable, and is very fundamental. And I told Fermi about it. He said: "it is interesting, you go and develop it." I thought for a couple of days, and could not go on. From this episode, I draw a lesson that everyone has his or her own line of thinking. If you have an original idea, it is worth thinking about, but do not be obsessed with it. If you are unsuccessful after a couple of days, you should move on to another problem. Always thinking about one thing can make one crazy. Another story is as follows. I had an idea, when I was a graduate student. Noting that the Maxwell equations have a close relation with the conservation of electric charge, and given that conservation of the isotopic spin had been confirmed by experiment, couldn't there be another kind of gauge field? I developed this idea for a couple of days, and could not go on. After half a year, one year, I felt this idea is very good, so I went back to try. I tried for several times, until 1954. Then the importance of this problem became clearer. To use gauge fields to write down the interaction is a principle, at least for a class of interactions. So I again tried to develop my idea. At that time I was sharing an office with Mills at the Bookhaven Laboratory. We added an additional term $[A_\mu, A_\nu]$ on the right-hand side of $F_{\mu\nu} = \partial_\mu A_\nu - \partial_\nu A_\mu$, where the A_μ are 2×2 matrices, and the difficulty was finally overcome. Non-Abelian gauge theory was thus born. The lesson from this story is that, if you have an original idea, do not give it up easily, but do not stick to it indefinitely; you should also pay attention to other things, and broaden your perspective. This is like playing go, if you have a disadvantage at some area, do not stick to it indefinitely: change to another area to develop a territory. Afterwards, circumstances change, and perhaps the old area becomes alive. So there are two points, not only do not give up your original idea, but also work on other things for some time. As to your biological field, I do not understand this; you may discuss it with others.

Q: What do you think of going abroad for graduate study with self-support?

C. N. Yang: My thought is, there are many many people in China, studying abroad with self-support is nothing bad. Now that we are discussing this topic, let me elaborate more. I have two comments. First, there are many people from Chinese mainland going abroad for graduate study, and many of them did not go back after receiving their doctoral degrees. This was anticipated. Many years ago, I said it in China. Now many people care about it, regarding it as a great loss, and very bad. I do not agree to it. Many people remain abroad, but there are more people who haven't gone abroad. So it is not a big problem. In 1950s to 1970s, among those from Taiwan studying abroad, those going back comprise less than 1 to 2 percent. But this did not stop the economic development of Taiwan. Taiwan's economy has its shortcomings, but it has developed a lot during that period. Second, let me illustrate matters by using a story. Three weeks ago, I was taking a taxi in Hong Kong. The driver was a woman. I said: "You speak good Mandarin." She said that she went to middle school in Beijing in the 1950s. She asked what I do. I told her that I am a physics professor living in America. She said very good, but she had a question to ask me. She has a son who graduated from a middle school in Guangzhou and entered the Physics Department of Fudan University. His classmates all wanted to go abroad. So does he. But to go abroad through CUSPEA (China-U.S. Physics

Examination and Application), there are only seven or eight successful students each year. He was a top student when he was in middle school, but the competition is strong in Fudan, and he felt that it was not hopeful for him to go abroad through CUSPEA. So he worried very much. Through an American organization, he made contact with a university (I have not heard of its name), but 70 to 80 thousand Hong Kong dollars are needed. She asked me what she should do. After arriving at the destination, I wrote a note to his son. I wrote: "My name is Chen Ning Yang, and I have two points. One, Fudan University is a first class university, the undergraduate physics education you are receiving in Fudan can only be better, not at all worse, than the undergraduate education in any American university. Second, after your graduation from Fudan, many American universities will accept you for graduate study and provide a teaching assistantship, no matter whether you are sent by the government or go by yourself, since Fudan has a very good reputation in America. I think this is the best way for you to study physics. Finding an arbitrary university in a hurry does nothing good for you."

Acknowledgment

We thank Prof. C. N. Yang for his advice.

References

1. C. N. Yang, *Proceedings of the Graduate School of the University of Science and Technology of China* (in Chinese), October issue (1986), reprinted in *Collected Papers of C. N. Yang* (in Chinese) (East China Normal University Press, Shanghai, 1998).

https://doi.org/10.1142/9789819826636_0008

China should not build a supercollider at this time

C. N. Yang
Tsinghua University, China

Published 26 October 2016

On August 29, a WeChat Public Platform "Mr. Gu talks about geometry" (ID: conformalgeometry) posted an article entitled "Shing-Tung Yau: Some opinions on China's construction of high energy colliders and reply to questions in the media," in which Yau mentioned that he supports the proposal for China to build a supercollider, while I (Yang) oppose the proposal, and he found my position hard for him to believe. One paragraph in that article reads as follows:

> "The fundamental theories behind these experiments are based on Professor Yang's theory. After each breakthrough, we admire Professor Yang's deep insights even more! Therefore it is perplexing for me to hear that Professor Yang has objection to the further development of high-energy physics."

Professor Yau misunderstood my position! I have absolutely no objection to the continued development of high-energy physics. What I am opposed to is the proposal for China to build a supercollider now, for the following reasons:

(1) The United States had a painful experience about building such a supercollider: In 1989, the United States initiated the construction of the world's largest collider with a budget of US$3 Billion. The budget was revised upwards several times, reaching US$8 Billion and arousing a lot of oppositions, until in 1992 the US Congress painfully terminated the project, writing off US$3 Billion already spent. This experience has led to a common belief that trying to build a large collider is like falling into a bottomless pit.

Currently the world's largest collider is the Large Hadron Collider (LHC) at CERN. In 2012, 6,000 physicists working with this collider discovered the Higgs particle, which validates the Standard Model — a great contribution to particle physics. It took many years to build the LHC and the total cost of construction plus detectors is more than US$10 Billion. The budget for the supercollider by the Institute of High Energy Physics (IHEP) of the Chinese Academy of Sciences (CAS) cannot be less than US$20 Billion.

Reproduced from International Journal of Modern Physics A, Vol. 31, No. 30, 1630053 (2016), DOI: 10.1142/S0217751X16300532

(2) IHEP proposes that the supercollider should be constructed in China with the cost shared among many countries. China's share of the cost must be considerable. Today the whole world is amazed that China's GDP has leapt to second in the world. But China is still a developing country with a per capita GDP less than that of Brazil, Mexico or Malaysia. There are hundreds of millions of farmers and migrant workers and there are many urgent unsolved environmental problems, education problems, medical and health problems, etc. The construction of a supercollider, with such an astonishing cost, would be detrimental to solving these urgent problems. Hence I think at present it should not be considered.

(3) The construction of the supercollider would significantly squeeze the funding for other basic sciences, including life sciences, condensed matter physics, astrophysics and so on.

(4) Why do many high-energy physicists actively support the construction of a supercollider? The reasons are as follows:

> **(A)** High-energy physics is a novel field that emerged after World War II. In the last 70 years it has brilliant achievements. The validation of the Standard Model has brought us in-depth understanding about three basic types of forces in nature. Nevertheless there are two major unsolved problems: **(I)** There are fundamental difficulties in understanding the remaining fourth force — gravitation; **(II)** We still do not yet know how to unify force and mass. It is, of course, the desire of all physicists to solve these two problems.
>
> **(B)** Some high-energy physicists hope to use the supercollider to discover "supersymmetric particles," and hence shed light on these two problems. The search for supersymmetric particles, however, has been going on for many years, and all to no avail. Today, it is just a conjecture of some high-energy physicists that a supercollider might be able to find supersymmetric particles. Most physicists, including myself, believe that the existence of supersymmetric particles is merely a conjecture without any experimental basis. Hoping to use the supercollider to find those conjectured particles is no more than a conjecture upon a conjecture.

(5) Have the great achievements of high energy physics in the last 70 years benefited the life of humankind? No. If the supercollider proposed by IHEP could be built and really succeeds in pushing high-energy physics a big step forward, would the life of humankind be benefited? I do not think that is going to happen in the short term or medium term, say 30 years or 50 years. And I know that the vast majority of physicists agree with me.

(6) It has been more than 30 years since the IHEP was established. How do we evaluate its achievements in these 30 years? Today, among the world's leading high-energy physicists, China accounts for only one or two percent. For the construction of the large collider, its design and operations and analysis after its completion, will thus be led 90% by non-Chinese scientists. If it finally leads to a Nobel Prize, would the winners be Chinese?

(7) Would high-energy physics come to an end if we do not build a supercollider? I don't think that would be the case. There are at least two directions worth exploring: **(A)** Finding new accelerator principles; **(B)** Looking for beautiful geometric structures, as string theorists do. Researches along these two directions are not so costly, and would be in line with the overall trend of global economic development.

https://doi.org/10.1142/9789819826636_0009

Jeremy Bernstein's monologue

C. N. Yang
Institute for Advanced Study, Tsinghua University, P. R. China

Published 30 May 2017

There appeared recently on the web a monologue[a] by Jeremy Bernstein (1929–) about how he had written his famous *New Yorker* article in 1962. The following is a transcription of the monologue: Since the recording on the web was not of good quality, the transcription probably contains mistakes.

"*Still had no idea how to do this. But I was going back to CERN in Geneva for the summers. Now about a year had gone by, I hadn't come up with anything. But life works in a strange way. I played a lot of tennis, and I sprained the ankle. And I lived in a building in which T. D. and his wife lived. And they took sympathy with me. I drove back and forth with him to CERN every day from where we were living. And I talked to him and I got to know him. And I thought, you know, I think what I can do is, I can do a profile of Lee and Yang. So I said*: "*can I do this?*" *He was not enthusiastic, but was not totally negative. So I went back to do a profile of Lee and Yang. And Shawn edited it actually. . . .*

I forget exactly how I had done that. When I referred to things, I think I may have called them Lee and Yang. I don't know, anyway. There was a strange thing, because when thinking back, Lee and Yang were changed occasionally to Yang and Lee, it was kind of odd. So Shawn called me and said, you know, they have changed from Lee and Yang to Yang and Lee in various places, do you have any idea why? And I said "No, I don't know why." It turned out that they had a terminal fight and broke up. And, some people blamed me, but, you know, I didn't do anything. I think Dyson blamed me. But I certainly just wrote this profile and didn't do anything. And then the next summer I had to talk with T. D. about it. He was very disturbed. But I think what happened often in the collaborations is that, when the collaboration started, Lee was younger and a junior person, Yang was older and from a different social class in China. And in the course of things, I think that, Lee began getting most of the ideas and Yang was getting most of the credit. I think that was the

[a]Web of Stories, https://www.webofstories.com/play/jeremy.bernstein/1.

Reproduced from Modern Physics Letters A, Vol. 32, No. 19, 1730017 (2017), DOI: 10.1142/S0217732317300178

source of their tension. I've seen this before. I saw this with Gell-Mann and Pais. I say it was a canonical thing that happened in these collaborations. So I felt very very badly about it. I felt that ... I just really felt very badly that this had happened, and I had some responsibility for it.

Lee left the Institute and went back to Columbia, which was good for them. I got a note from Dyson, saying once we forgive you but twice we won't, which was very disturbing. And then I tried to do a profile of Dirac, I had a one-day interview. I think Oppenheimer got hold of it and told me not to do it. It was a pity because I could have done a very nice profile of Dirac...."

My comment: This monologue in a way is a confession by Bernstein in his old age. He was quite confused, splicing real and imagined events from different periods together. But the main theme is clear: Now he "*felt very badly about*" the publication of the article, for "*I had some responsibility for it.*"

A scientific collaboration is built on contributions by several individuals, each with his/her special talent and special experience. The more successful the collaboration is, the more trust and consideration are required to keep the collaboration going. Media probing of the intimate details of successful scientific collaborations are potentially very destructive. Oppenheimer, I, and several friends knew this in 1962, and had tried to stop the publication of the article, but without success.

There was, in between WWI and WWII, a very successful collaboration between two powerful British mathematicians, Hardy (1877–1947) and Littlewood (1885–1977). They were entirely different persons: different in personal character, different in style of research. Yet they produced beautiful mathematics together for almost 30 years. Many people were, of course, interested in how they could have done so. C. P. Snow was one of them. He was in addition a close friend of Hardy's. In a very perceptive passage[b] he revealed the fundamental secret about that famous collaboration:

"*Hardy talked to me, over a period of many years, on almost every conceivable subject, except the collaboration. He said, of course, that it had been the major fortune of his creative career: He spoke of Littlewood in the terms I have given; but he never gave a hint of their procedures. I did not know enough mathematics to understand their papers, but I picked up some of their language. If he had let slip anything about their methods, I don't think I should have missed it. I am fairly certain that the secrecy — quite uncharacteristic of him in matters which to most would seem more intimate — was deliberate.*"

I wonder, if this passage had been written before 1962, and if Bernstein had read it and had learned deeply of Hardy and Littlewood wisdom, would he realize that he should not meddle into the very successful Lee–Yang collaboration?

[b]Foreword by C. P. Snow, in *A Mathematician's Apology*, G. H. Hardy (1967).

Reflections on His Peers

https://doi.org/10.1142/9789819826636_0010

C. Y. CHAO, PAIR CREATION AND PAIR ANNIHILATION

BING AN LI*

Physics Department, University of Kentucky, Lexington, Kentucky 40506, USA

and

C. N. YANG

Institute for Theoretical Physics, State University of New York at Stony Brook, Stony Brook, New York 11794-3840, USA

Received 7 June 1989

C. Y. Chao's contribution to physicists' acceptance of QED in 1933–1934 through his experiments of 1930 is analyzed. It is pointed out that Blackett and Occhialini's key suggestion of 1933 about hole theory was based on identifying Chao's "additional scattered rays" (1930) as due to pair annihilation.

1. Introduction

The Dirac theory[1] of the infinite sea of electrons, a milestone in the history of twentieth century physics, is now universally accepted as an integral part of the foundation of particle physics. Historically, however, it had a very difficult time of being accepted.[2,3] What eventually reversed the tide of disbelief was the 1932 discovery of the positron,[4] and the subsequent understanding[5] of the processes of pair creation and pair annihilation.

Actually these latter processes had been discovered experimentally a few years earlier, but had not been theoretically understood. These earlier discoveries were reported in

(a) papers independently submitted in the month of May, 1930, by three groups of physicists, Tarrant[6] in Cambridge, England, Meitner and Hupfeld[7] in Berlin-Dahlem and C. Y. Chao[8] in Pasadena, describing the discovery of the "anomalous absorption" of ThC″ γ-rays (2.65 MeV) by heavy elements, and

(b) a paper of Chao[9] in late 1930 on a new experiment in which he discovered "additional scattered rays" of ThC″ γ-rays in Pb.

Papers (a), in retrospect, represented the first observation of the pair creation process and paper (b) the first observation of the pair annihilation process. In the ensuing two years, 1931–1932, the "anomalous absorption" and "additional scattered rays" attracted much attention among theoretical physicists, and stimulated important further experimental research. To appreciate the impact of Chao's papers, we quote here from a 1983 article[10] by C. D. Anderson:

> *My thesis work as a graduate student at Caltech consisted of studying, by means of a Wilson cloud chamber, the space distributions in various gases of*

* Permanent address: Graduate School, Academy of Sciences, Beijing, P R China.

Reproduced from International Journal of Modern Physics A, Vol. 04, No. 17, pp. 4325–4335 (1989), DOI: 10.1142/S0217751X89001813

Fig. 1. C. Y. Chao in 1927. This photograph was taken just before he came to the USA for graduate education.

photoelectrons produced by x rays. At the time (1927–30) that I was doing this work, Dr. Chung-Yao Chao, working in a room close to mine, was using an electroscope to measure the absorption and scattering of γ rays from ThC". His findings interested me greatly. At that time it was generally believed that the absorption of "high-energy" γ rays (2.6 MeV from ThC") was almost wholly by Compton collisions, as governed by the Klein-Nishina formula. Dr. Chao's results showed clearly that both the absorption and scattering were substantially greater than calculated by the Klein-Nishina formula. A detailed explanation of these anomalous effects was not possible from his experiments because one could not obtain detailed information from an electroscope. My proposed experiment was to study the interaction of the ThC" γ rays with matter using a cloud chamber operated in a magnetic field: to view the secondary electrons produced in a thin lead plate inserted in the cloud chamber, to measure their energy distribution, and to see what further light could be thrown on Chao's results.

Also from an article by Hayakawa[11] about his conversation in 1980 with Occhialini:

Occhialini spoke highly of Chao's achievements and mentioned how his study of the anomalous absorption of ThC" gamma-rays triggered off their related research even in England.

Thus today in the 1980's Anderson and Occhialini both emphasize Chao's work as having stimulated in the early 1930's their own dramatic research which were instrumental in transforming physicists' understanding of quantum electrodynamics. Neither mentioned Chao's competitors' work in this connection.

In the present paper we trace the history of these discoveries of the "anomalous absorption" and the "additional scattered rays" in 1930, the essential role that Chao had played in them, and how his work was important for Blackett and Occhiliani's later suggestion about pair annihilation.

2. Background

In the years around 1930 many important events happened in physics. This was an extremely active and exciting, but also confusing period.

One of the great confusions was the question whether electrons and protons are the constituents of nuclei. Since these were the only two elementary particles known at that time, (excluding the photon), it was natural to assume that nuclei were made of them. However, this assumption encountered[12,13] many grave difficulties. Another confusion arose because β-decay spectra seemed to be continuous and that had led Bohr and others to propose[13] that energy may not be conserved in β-decay.

On a more theoretical level, Dirac's equation and the hole theory were producing great anguish and soul searching. Most distinguished physicists were against it.[14] Of course, everyone had to acknowledge the brilliant achievement of Dirac in explaining, out of sheer originality and simple arithmetic, the spin and the magnetic moment of the electron. But the negative sea was universally regarded as a sickness. Pauli· had said[14]

> *Any theory which had such a sickness must agree with experience only by accident.*

Against such a background there was great interest in studying experimentally elementary scattering processes involving the electron such as the Compton scattering, and using that to check theoretical calculations. There were at that time three different formulae for Compton scattering:

(a) *The Compton formula*[15]

Compton modified J. J. Thomson's classical theory, to account for the shifted wavelength and the recoil effect. He obtained for the cross section:

$$\sigma = \frac{8\pi}{3} \frac{e^4}{m^2c^4} \frac{1}{1 + 2\alpha}, \tag{1}$$

where $\alpha = h\nu/mc^2$, which is different from Thomson's result. Compton's theory was also different from J. J. Thomson's classical theory in another way: for hard γ rays, the outgoing wave was concentrated in the forward direction. From the point of view of quantum theory, as pointed out by Compton himself, his was not a correct theory.

(b) *The Dirac*[16] *and Gordon*[17] *formula*

Dirac and Gordon started from quantum mechanics and used different methods, arriving at the same formula

$$\sigma = \frac{2\pi e^4}{m^2c^4}\frac{1+\alpha}{\alpha^2}\left[\frac{2(1+\alpha)}{1+2\alpha} - \frac{1}{\alpha}\ln(1+2\alpha)\right]. \tag{2}$$

The spin was not taken into account.

(c) *The Klein-Nishina formula*[18]

$$\sigma = \frac{2\pi e^4}{m^2c^4}\left\{\frac{1+\alpha}{\alpha^2}\left[\frac{2(1+\alpha)}{1+2\alpha} - \frac{1}{\alpha}\ln(1+2\alpha)\right] + \frac{1}{2\alpha}\ln(1+2\alpha) - \frac{1+3\alpha}{(1+2\alpha)^2}\right\}. \tag{3}$$

This formula was derived by grafting onto classical radiation theory Dirac's relativistic wave equation. After Dirac's hole theory,[1] Dirac[19] and Waller[20] showed that in the hole theory a correct second order perturbation calculation gives the same result[21] as the Klein-Nishina formula.

Formulae (1), (2) and (3) have identical low energy limits. They became different at high energies, i.e. with "hard γ-rays". Thus in 1929–1930, to check and differentiate between these theories, three experiments[6–8] were performed to measure the absorption coefficients of hard γ-rays, by Chao in Pasadena, Tarrant in Cambridge, England and by Meitner and Hupfeld in Berlin-Dahlem.

3. Anomalous Absorption

C. Y. Chao was born in Zhejiang Province, China, in 1902. After graduation from the Chemistry Department of the South-East University in 1925, he became a teaching assistant of Chi-Sun Yeh in Tsing Hua University. In the summer of 1927 Chao came to the United States and became a graduate student of R. A. Millikan at the California Institute of Technology. Millikan asked Chao to measure the absorption coefficients of hard γ-rays in various materials to test the Klein-Nishina formula (3). According to Chao in 1986,[22] Millikan had been initially inclined to believe that (2) and not the Klein-Nishina formula (3) was consistent with cosmic ray data.

Near the end of 1929, Chao completed his experiments. He found that for light elements the results agreed with the Klein-Nishina formula, but that for heavy elements, for instance Pb, the experimental values of the absorption coefficient were greater than that given by the formula. Millikan at first did not believe Chao's results. Chao's paper was thus not submitted for publication for several months. Fortunately, Professor L. S. Bowen knew Chao's experiments in detail. He convinced Millikan that Chao's experiments were reliable, and suggested to send Chao's paper for publication as soon as possible.[22] The submission date[8] was May 15, 1930. Two weeks before that, on April 29, 1930 Chao's results had been presented[23] at the U.S. National Academy of Sciences.

Tarrant[6] and Meitner and Hupfeld[7] reached the same main conclusions as Chao, independently, but there were some differences in detail:

(a) The dependence of the absorption coefficient on the atomic number of the medium was irregular in Tarrant's experiment[24] and had a "jump" in Meitner and Hupfeld's[26] (Fig. 2). They led to controversies. Chao's result was, in contrast, quite smooth and did not lead to controversies.

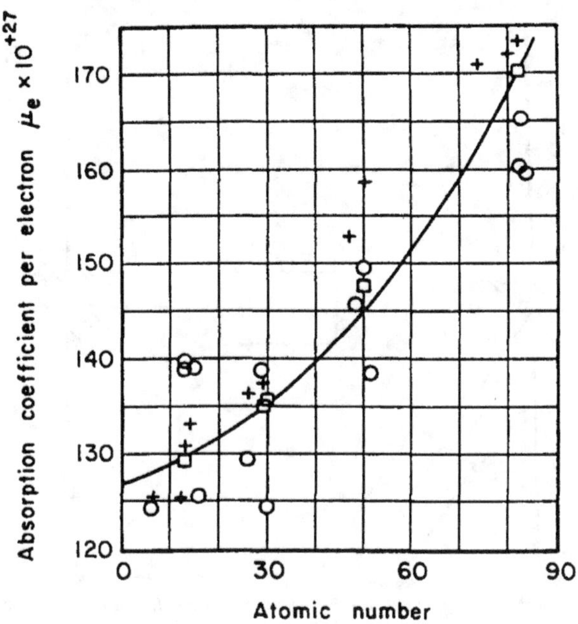

Fig. 2. Comparison of the Three Experiments in 1930 on Anomalous Absorption. The three experiments are designated as

○ Tarrant[6] (1930)

\+ Meitner and Hupfield[7] (1930)

□ Chao[8] (1930)

The graph is redrawn from a graph in Tarrant [*Proc. Roy. Soc. London* **A135** (1932) 223] from which the data points of Jacobson and those of Tarrant's later experiments are deleted. The data points of Tarrant[6] (1930) are added by us. The curve, drawn by Tarrant in 1932, *fits beautifully* Chao's 1930 data.

(b) The detectors used were quite different. According to Chao in 1986,[22] his detector, a pressurized ionization chamber at 25 atmospheres and a vacuum electroscope, was the more reliable. Figure 2 demonstrates that this is in fact an understatement.

The additional absorption by heavy elements of hard γ-rays found by all three experiments was called the "Anomalous Absorption" or the "Meitner-Hupfeld effect", the latter terminology having originated from Meitner and Hupfeld's friends. All three publications[6–8] speculated that the anomalous absorption was due to some unknown nuclear effects.

4. "Additional Scattered Rays"

In order to obtain more information about the absorption mechanism of hard radiation in matter, Chao started, soon after his first experiment, a new one to study the intensity and angular distribution of the scattered radiation (Fig. 3). This was a difficult experiment since the scattered radiation was weaker than the background. The experiment was

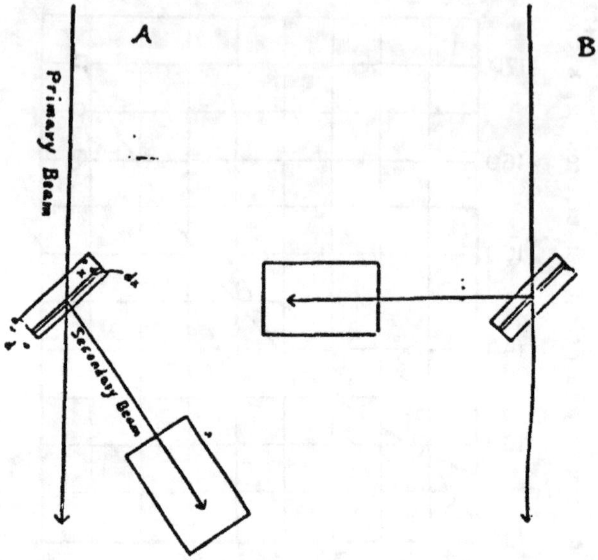

Fig. 3. Diagram of Second Experiment of Chao. From *Science Reports of Tsing Hua University* **1** (1932) 159.

published[9] in 1930. It was not until one year later that other experimental groups joined in this effort to study the scattered radiation. Furthermore these later efforts turned out to be unclean and inconclusive, creating controversies that distracted theorists' attention and unfortunately lessened the impact of Chao's results. We shall come back to these points below in Sec. 6.

What Chao found in his second experiment[9] were:

(a) Accompanying the anomalous absorption there is an additional "scattered" radiation from the heavy scatterer.

(b) Such additional scattered radiation is essentially isotropic.

(c) The wavelength of the additional scattered radiation was measured to be 22 X.U. which corresponds to photons at .5 MeV.

It is impressive how Chao was right on the dot in all of these conclusions. He has found pair annihilation! In this process:

$$e^+e^- \to \gamma\gamma\,, \tag{4}$$

each photon has an energy of approximately 1/2 MeV which is exactly what Chao found. However, nobody understood at the time the theoretical meaning of his findings until much later, as we shall discuss in the next section.

5. Interpretation

During the two and a half years from mid-1930 to the beginning of 1933, there were the following three topics before the physics community which according to later understanding are all parts of quantum electrodynamics (QED):

Anomalous Absorption
Additional Scattered Rays
Dirac's Theory

On Dirac's theory there were intensive discussions, including the calculations[27] of Oppenheimer, Tamm and Dirac on the cross section for the annihilation process (4), but none of these authors related this process to Chao's additional scattered rays. (This is very likely due to the fact that nuclear physics was in a very confused state. Thus quite naturally Chao's result and the anomalous absorption, both strange phenomena, were thought of as nuclear phenomena.)

On additional scattered rays we found no theoretical paper in this period. (See Sec. 6 below where it is shown how the final understanding in 1933 in terms of pair annihilation and pair creation were based on additional scattered rays and not on anomalous scattering.)

On anomalous absorption there was much theoretical discussion. Oppenheimer[28] tried the idea that it is due to the photoelectric effect but concluded that QED was wrong, a popular theme in those days. Heisenberg[29] and Gamow[30] speculated that it is due to some nuclear processes.

None of these theoretical works produced great progress. It was only after Anderson had discovered[4] the positron in September 1932, and Blackett and Occhialini[5] had obtained many positrons a few months later with their beautiful triggered cloud chamber, that the question was raised[5] about the property of the positron in its passage through matter. This led Blackett and Occhialini to Dirac's earlier calculation[27] of the annihilation process (4), and to their suggestion that Chao's additional scattered rays is the result of the annihilation process.

Blackett and Occhialini also speculated about the production of positrons by pair creation. They did not, however, arrive at the (correct) idea of a QED pair creation process in the Coulomb field of heavy nuclei. But within a few months, this idea was conceived and developed according to the Dirac theory (i.e. QED) by Oppenheimer and Plesset,[31] who found that the result agreed with the earlier experimental findings on anomalous absorption. Also Anderson observed pair creation by photons in a cloud chamber.[32]

After 1933, with all these clarifications, QED entered into a period in which there were great successes in all applications. The only disturbing things that remained were the divergencies.

6. Chao's Crucial Value of .5 MeV

The paper of Blackett and Occhialini[5] made a great impact not only because they reported the findings of many examples of the newly discovered positive electron, but also because they suggested that the "anomalous absorption" and the "additional scattered rays" were due to pair creation and pair annihilation respectively, which led to a turnaround in physicists' perception of the validity of Dirac's theory. The crucial passage on this matter in this paper was:

> *Perhaps the anomalous absorption of gamma-radiation* by heavy nuclei may be connected with the formation of positive electrons and the re-emitted radiation with their disappearance. The re-emitted radiation is, in fact, found experimentally to have an energy of the same order as that to be expected for the annihilation spectrum.* (A)
>
> * *Gray and Tarrant, 'Proc. Roy. Soc.' A, vol. 136, p. 662 (1932); Meitner and Hupfeld, 'Naturwiss.,' vol. 19, p. 775 (1931); Chao, 'Phys. Rev.,' vol. 36, p. 1519 (1931).* (B)

Passage (A) was great physics. Footnote (B) was careless history. In particular this footnote served C. Y. Chao very poorly. It contains two typographical or careless mistakes:

(i) Chao's article in the Physical Review was published in 1930, not 1931. It preceded the other two papers by one or two years.

(ii) All three cited papers were about additional scattered rays, not anomalous absorption, although the asterick in the text was attached to the latter.

More important, Blackett and Occhialini's argument really rested on Chao's paper alone, which fact was obscured by this indiscriminate footnote, as we shall now show.

Blackett and Occhialini arrived at the great suggestions (A) quoted above in a section of their paper entitled "*The Hypothetical Properties of the Positive Electron*". This section started with raising the question of why positive electrons "*have hitherto eluded observation*", followed by

> *It is clear that they can have only a limited life as free particles since they do not appear to be associated with matter under normal conditions. . . . it seems . . . likely that they disappear by reacting with a negative electron to form two or more quanta.*

They went on to say that this disappearance mechanism was given "*immediately by Dirac's theory of electrons*", and that they had talked to Dirac, who had shown them his paper of 1930 giving the annihilation cross-section. Blackett and Occhialini then concluded that Dirac's theory predicted a lifetime for the positive electron "*that is long enough for it to be observed in the cloud chamber but short enough to explain why it had not been discovered by other methods*", going on to say that it is possible however to observe the annihilation process because it gives rise to photons with a spectrum that peaks at .5 MeV. Quotation (A) then follows.

This whole series of arguments represent magnificent physics. They also show that the *focus* of Blackett and Occhialini's reasoning was *on the annihilation process*. This process was identified as giving rise to the "additional scattered rays" on the strength of the fact that the annihilation radiation has a spectrum peaked at .5 MeV (see above), in agreement with what was "*found experimentally*" [quote (A)]. Thus the matter of the energy of the additional scattered rays belong experimentally ~.5 MeV was *crucial* to their reasoning.

For some reason, Blackett and Occhialini did not mention,[33] that of the three papers they cited in quote (B) above (all on additional scattered rays,) *only Chao's* had given the *correct crucial value of .5 MeV*. Meitner and Hupfeld's paper (1931), one year after Chao's, did not find these additional rays at all. Gray and Tarrant's paper (1932), two years after Chao's, did find such an additional scattered ray at ~.47 MeV, but did also find a ~.92 MeV component which was very confusing and which refused to go away even in a later paper[34] of 1934.

Incidentally one may wonder why throughout Blackett and Occhialini's paper no references at all were made to the three papers of 1930 on anomalous scattering (which had attracted more attention than those on additional scattered rays because it was discovered earlier and was based on less difficult experiments. See Sec. 5). The answer is: They were not the focus of Blackett and Occhialini's arguments which were on the annihilation process. Furthermore, Blackett and Occhialini did not pinpoint the mechanism of pair creation as taking place *in a Coulomb field*, which was only later investigated theoretically by Oppenheimer and Plesset.[31]

Looking into the history[35] of the discovery of pair creation and pair annihilation, we are deeply impressed with Chao's experiments. They probed into important questions. They were difficult, as evidenced by the pitfalls that his competitors fell into both in the anomalous absorption experiment and in the additional scattered rays experiment. They had the classical beauty of simple, reliable experiments that stood the test of time.[36] Unfortunately because of a careless reference in Blackett and Occhialini's paper and because of the confusion and controversies generated by the other experiments, Chao's papers had not received the credit that they so richly deserve.

Acknowledgments

The authors wish to thank L. M. Brown and G. Ekspong for comments which had contributed to this work. They also acknowledge partial support of the NSF under grant PHY8507627.

References

1. P. A. M. Dirac, *Proc. Roy. Soc. London* **117** (1928) 610; **118** (1928) 351; **126** (1930) 360.
2. See the papers of D. F. Moyer (Ref. 14 below, p. 1059–1061), of L. M. Brown (Ref. 13 below) and of L. M. Brown and D. F. Moyer (Ref. 25 below).
3. See G. Gamow, *Thirty Years that Shook Physics* (Double Day, 1966) for jokes about Dirac's "Donkey electron" at Copenhagen in 1930–1932.
4. C. D. Anderson, *Science* **76** (1932) 238.
5. P. M. S. Blackett and G. P. S. Occhialini, *Proc. Roy. Soc.* **A139** (1933) 699.
6. G. T. P. Tarrant, *Proc. Roy. Soc.* **A128** (1930) 345.
7. L. Meitner and H. H. Hupfeld, *Naturwiss* **18** (1930) 534, detailed report of this experiment was in *Z. Phys.* **67** (1931) 147.
8. C. Y. Chao, *Proc. Nat. Acad. of Sci.* **16** (1930) 431, a fuller report of this experiment was in *Science Reports of Tsing Hua Univ.* **1** (1932) 159.

9. C. Y. Chao, *Phys. Rev.* **36** (1930) 1519, a fuller report of this experiment was in *Science Reports of Tsing Hua Univ.* **1** (1932) 159.
10. C. D. Anderson and H. L. Anderson, *The Birth of Particle Physics*, eds. L. Brown and L. Hadsson (Cambridge, 1983) p. 131.
11. S. Hayakawa, *Shizen (Nature)* September 1980, p. 75. Translation in letter by S. Hayakawa to C. N. Yang dated July 4, 1985.
12. H. A. Bethe, *Proceedings of a Symposium on the 1930's Nuclear Physics in Retrospect*, ed. R. H. Stuewer (University of Minnesota, 1977), p. 11 and E. P. Wigner, *ibid.* p. 157.
13. L. M. Brown, *Phys. Today* **31** (1978) 23.
14. D. F. Moyer, *Am. J. Phys.* **49** (1981) 1055.
15. A. H. Compton, *Phys. Rev.* **21** (1923) 483.
16. P. A. M. Dirac, *Proc. Roy. Soc. Ser.* **A111** (1926) 405.
17. W. Gordon, *Z. Phys.* **40** (1926) 117.
18. O. Klein and Y. Nishina, *Z. Phys.* **52** (1929) 853.
19. P. A. M. Dirac, *Proc. Camb. Phil. Soc.* **26** (1930) 361.
20. V. I. Waller, *Z. Phys.* **61** (1930) 837.
21. It is interesting to note that even after the discovery of the positron in 1932 Pauli was still against the hole theory. In June 1933 Pauli submitted a paper, later published in the *Helv. Phys. Acta*, on the scattering of photons by a moving electron, in which he did not use the concept of the infinite sea, but he did quote Dirac[19] and Waller.[20] According to Moyer, *Am. J. Phys.* **49** (1981) 1123, in lectures in 1935–1936, Pauli called his work with Weisskopf on the scalar meson the "Anti-Dirac Theory". Pauli in fact gave explicit resons for his dislike of the hole theory which were quoted in length by Moyer.
22. Our interview with Prof. Chao in his apartment in Beijing, June 2, 1986.
23. C. Y. Chao, *Proc. Nat. Acad. Sci.* **XVI** (1930) 421.
24. See Ref. 6 above and G. T. P. Tarrant, *Proc. Roy. Soc.* **A135** (1932) 223. See also Ref. 25.
25. L. M. Brown and D. F. Moyer, *Am. J. Phys.* **52** (1984) 130.
26. See Ref. 7 and Refs. 24, 25.
27. J. R. Oppenheimer, *Phys. Rev.* **35** (1930) 562, 939; *Nature* **125** (1930) 616; I. Tamm, *Z. Phys.* **62** (1930) 545; P. A. M. Dirac, *Proc. Camb. Philos. Soc.* **26** (1930) 361.
28. J. R. Oppenheimer, *Phys. Rev.* **38** (1931) 57.
29. W. Heisenberg, *Z. Phys.* **77** (1932) 1; **78** (1932) 156; **80** (1933) 587.
30. G. Gamov, *Constitution of Atomic Nuclei and Radioactivity* (Clarendon, Oxford, 1931) p. 79–82.
31. J. R. Oppenheimer and M. S. Plesset, *Phys. Rev.* **44** (1933) 53.
32. C. D. Anderson, *Science* **77** (1933) 432.
33. Compare the Hayakawa statement about Occhialini's 1980 conversation quoted above in Sec. 1.
34. L. H. Gray and G. T. P. Tarrant, *Proc. Roy. Soc.* **A143** (1934) 681, 705.
35. See J. D. Stranathan, *The "Particle" of Modern Physics*, (Blackiston, 1942), and R. A. Millikan, *Electrons, Positive and Negative, Protons, Photons, Neutrons, Mesotrons and Cosmic Rays* (University of Chicago Press, 1947) for views in the 1940's of these developments.
36. Chao performed a third experiment in Germany in 1931 on his way home to China, on the dependence of the anomalous absorption on the wave length of the incident beam: *Proc. Roy. Soc.* **A135** (1932) 206. In 1933 he published a fourth experiment with T. T. Kung performed at the National Tsing Hua University in China: *Nature* **132** (1933) 709. This article was followed by an interesting note of Rutherford:

    It is obvious from a letter to me which accompanied the above communication that Prof. Chao and Mr. Kung have not yet heard of the recent work concerning the positive electron, and in particular of the creation of a pair of electrons, a negative and a *positive, by the conversion of a γ-ray of high energy in the strong electric field of a nucleus. The experiments they describe provide valuable additional evidence of this phenomenon, and would doubtless have been interpreted by them in this way rather than as a nuclear disintegration. It is interesting to note that the magnitude of the effect is about the same as is found in other experiments.*

https://doi.org/10.1142/9789819826636_0011

ALBERT EINSTEIN: OPPORTUNITY AND PERCEPTION*

CHEN NING YANG
Huang Ji-Bei & Lu Kai-Qun Professor
Tsinghua University, Beijing 100084, China
and
The Chinese University of Hong Kong, Shatin, Hong Kong

I

The year 1905 has been called Albert Einstein's "Annus Mirabilis." It was during that year that he caused revolutionary changes in man's primordial concepts about the physical world: space, time, energy, light and matter. How could a 26-year-old clerk, previously unknown, cause such profound conceptual changes, and thereby open the door to the era of modern scientific technological world? No one, of course, can answer that question. But one can, perhaps, analyze some factors that were essential to his stepping into such a historic role.

First of all, Einstein was extraordinary lucky: he was born at the right time, and was at the peak of his creative powers when the world of physics was shuddering from multiple crises. In other words, there was the lucky opportunity for him to change the course of physics, an opportunity unmatched, perhaps, since the time of Newton. Such lucky opportunities occur very very infrequently. In E. T. Bell's *Man of Mathematics*, Lagrange (1736–1813) was quoted as having said:

> *Newton was assuredly the man of genius par excellence, but we must agree that he was also the luckiest: one finds only once the system of the world to be established.*

Here Lagrange was referring to the words in Newton's introduction to the third and final volumes of his great *Principia Mathematica*.

> *I now demonstrate the frame of the system of the world.*

*Talk given at The 8th International Symposium on Foundations of Quantum Mechanics in the Light of New Technology (ISQM-Tokyo '05), 22–25 August 2005, Japan.

Reproduced from International Journal of Modern Physics A, Vol. 21, No. 15, pp. 3031–3038 (2006), DOI: 10.1142/S0217751X06033192

Fig. 1. Einstein as a Swiss patent clerk in 1905 when he revolutionized fundamental physics through the creation of the special theory of relativity.

While Lagrange was obviously envious of Newton's lucky opportunity, we detect little sentiment of a similar nature in what Einstein had publicly said of Newton:

> *Fortunate Newton, happy childhood of science ... In one person he combined the experimenter, the theorist, the mechanic and, not least, the artist in exposition. He stands before us strong, certain, and alone ...*

Turning to Einstein's own times, he had the opportunity to amend the system created by Newton more than 200 years ago. This lucky opportunity was of course open also to all scientists of his time. Indeed electrodynamics in a moving system had been a subject of intense discussions since the Michaelson–Morley experiment, first performed in 1881, repeated with greater precision in 1887. Amazingly Einstein was already intensely interested in this topic while still a student in Zurich. He had written to his future wife Mileva in 1899:

> *I returned the Helmholtz's volume and am now rereading Hertz's propagation of electric force with great care because I didn't understand Helmholtz's treatise on the principle of least action in electrodynamics. I'm convinced more and more that the electrodynamics of moving bodies as it is presented today doesn't correspond to reality, and that it will be possible to present it in a simpler way.*
>
> [From Albert Einstein/Mileva Marić, *The love letters*, Edited by Renn & Schulmann, Translated by Smith.]

The search for this simpler way led, six years later, to special relativity.

Many other scientists were also deeply interested in the subject. Poincaré (1854–1912), one of the two towering mathematicians at the time, was actively working on the same problem. Indeed the name *relativity* was not invented by Einstein. It was invented by Poincaré. One reads in his speech delivered one year before 1905 (in *Physics for a New Century*, AIP publication on History, Vol. 5, 1986):

> *The principle of relativity, according to which the laws of physical phenomena should be the same, whether for an observer fixed, or for an observer carried along in a uniform movement of translation; so that we have not and could not have any means of discerning whether or not we are carried along in such a motion.*

This paragraph not only introduced the term "relativity," but showed amazing insight which is absolutely correct *philosophically*. However, Poincaré did not understand the full implication in *physics* of this paragraph: later paragraphs in the same speech showed that he failed to grasp the crucial and revolutionary idea of the *relativity of simultaneity*.

Einstein was also not the first to write down the great transformation formula:

$$x' = \gamma(x - vt)\,, \quad y' = y\,, \quad z' = z\,,$$

$$t' = \gamma\left(t - \frac{vx}{c^2}\right), \quad \gamma = \frac{1}{\sqrt{1 - v^2/c^2}}$$

which had already been given by Lorentz (1853–1928), after whom it was, and still is, named. But Lorentz also failed to grasp the revolutionary idea of the relativity of simultaneity. He wrote later in 1915:

> *The chief cause of my failure was my clinging to the idea that only the variable t can be considered as the true time and that my local time t′ must be regarded as no more than an auxiliary mathematical quantity.*
>
> [cf. Pais' biography of Einstein, p. 167]

That is, Lorentz had the *mathematics*, but not the *physics*, and Poincaré had the *philosophy*, but also not the *physics*. It was the 26-year-old Einstein who dared to question mankind's primordial concept about time, and insisted that *simultaneity is relative*, thereby opening the door to the new physics of the microscopic world.

Almost all physicists today agree that it was Einstein who had created special relativity. Is that fair to Poincaré and Lorentz? To discuss this question let us quote from A. N. Whitehead [in *The Organization of Thought*, (Greenwood Press, 1974), p. 127]:

> *To come very near to a true theory and to grasp its precise application, are two very different things, as the history of science teaches us. Everything of importance has been said before by somebody who did not discover it.*

Lorentz and Poincaré indeed did seize the lucky opportunity of the time, and had worked very hard on one of the main problems, the electrodynamics in a moving system. But they both missed the crucial key point. They missed because they had "clung" to old concepts, as Lorentz himself later had said. Einstein did not miss because he had a **freer perception** of the meaning of space–time.

To have a **free perception**, one must simultaneously be close to the subject under investigation, and yet be able to examine it at a **distance**. Indeed the often used term **distant perception** shows the necessity of maintaining a certain distance in any penetrating discernment. But distant perception alone is not enough. It must be matched by a detailed close-up understanding of the problem at hand. It is the ability to freely adjust, assess and compare the close-up and distant views that constitutes **free perception**. Pursuing this metaphor, we might say that Lorentz had failed because he had *only* a close-up view, while Poincaré had failed because he had *only* a distant perception.

The great Chinese esthetician 朱光潜 (1897–1986) had emphasized the importance of "psychical distance" in artistic and literary creativity. I think that idea is very much the same as the distant perception discussed above, but in another area of intellectual activity. In the brilliant definitive scientific biography of Einstein, *Subtle is the Lord*, by Pais, the author chose one word to describe Einstein's character: **apartness**, and quoted at the beginning of Chapter 3:

> **Apart** ... *4. Away from others in action or function; separately, independently, individually.*
>
> [Oxford English Dictionary]

Indeed, **apartness, distance**, and **free perception** are related concepts, referring to an essential element in all human creativity, in science, in art, and in literature.

Another historic achievement of Einstein's in 1905 was his paper "*On a heuristic point of view concerning the generation and conversion of light*" written in March of that year. Historically this paper launched the revolutionary idea of light as quanta with discrete energy $h\nu$. The constant h had already been introduced by Planck in 1900 in his bold theoretical study of black body radiation. In subsequent years, however, Planck got cold feet, and began to hedge. In stepped Einstein in 1905, who not only did not hedge, but pushed forward courageously with his "heuristic point of view" of light quanta. That this courageous push was not generally appreciated can be gathered from the following sentences in a document written by Planck, Nernst, Rubens and Warburg, eight years later in 1913, when they proposed Einstein for membership in the prestigious Prussian Academy:

> *In sum, one can say that there is hardly one among the great problems in which modern physics is so rich to which Einstein has not made a remarkable contribution. That he may sometimes have missed the target in his speculations, as, for example, in his hypothesis of light-quanta, cannot really*

> *be held too much against him, for it is not possible to introduce really new ideas even in the most exact science without sometimes taking a risk.*
>
> [From Pais, *Subtle is the Lord*, 1982, p. 382]

Here the ridiculed "*hypothesis of light-quanta*" referred to Einstein's bold proposal of 1905 mentioned above. Despite such general derision, Einstein pushed *further ahead*, and in papers of 1916–1917 established the value of the momentum of the light quantum, leading later to the epoch-making understanding of the Compton effect in 1924.

The history of the birth of the revolutionary idea of the light quanta can be summarized as follows:

1905 : Einstein's paper on $E = hv$

1916 : Einstein's paper on $P = E/c$

1924 : Compton effect

Throughout these years, before the Compton effect was established in 1924, Einstein was **alone** in his insightful perception, at a time when entrenched conviction about waves was sacred to the whole physics community.

Fig. 2. Einstein giving a lecture in 1922 at the College de France in Paris.

II

Between 1905 and 1924 Einstein's main research interests were focused on the general theory of relativity. As a scientific revolution general relativity is unique in the history of mankind. The grandeur of its conception, its beauty, its sweeping scope, its spawning the awesome science of cosmology, and the fact that it was conceived and executed by one single person, reminds me of the act of *creation* in the old testament. (And I wonder whether Einstein himself had thought of this comparison.)

Fig. 3. Einstein in his study in his home on Haberlandstraße in Berlin.

Of course, one would also naturally think of other scientific revolutions, such as Newton's *Principia*, special relativity, and quantum mechanics. Some differences: Newton's work had grandeur, had beauty, had sweeping scope. Yes. But he had before him the works of Galileo, of Kepler and of earlier mathematicians and philosophers. He was not alone at the time in searching for the law of gravity. Special relativity and quantum mechanics were both profound revolutions. But they were hot topics worked on by many people at the time. Neither was the creation of a single person. For general relativity, Einstein did not seize any opportunity. He created the opportunity. Alone, through deep perception, he conceived the problem and after seven or eight years of lonely struggle produced a new *system of the world* of unimaginable beauty. It was an act of pure *creation*.

III

General relativity represented the geometrization of the gravitational field. It quite naturally led to Einstein's push for the geometrization of the electromagnetic field. Thus was born his idea to formulate an overall geometrization of all forces of nature, a *unified field theory*, which gradually evolved into his main research effort during the latter part of his life. The last seminars that he gave, for example, were in 1949–1950 at the Institute for Advanced Study in Princeton, and the subject was his latest attempt to incorporate the field strengths $F_{\mu\nu}$ of the electromagnetic field into an unsymmetrical metric $g_{\mu\nu}$. This attempt, as well as his earlier attempts in the same direction, was also unsuccessful.

As a consequence of this lack of success, and also because of the fact that, starting in the late 1920's, he directed his attention almost exclusively to this search, neglecting such newly developing fields as solid state physics and nuclear physics, he was often criticized, even ridiculed. His devotion to the unified field theory was called an *obsession*. An example of this criticism is what I. I. Rabi (1898–1988) had said in 1979 at the Einstein centennial in Princeton:

> *When you think of Einstein's career from 1903 or 1902 on to 1917, it was an extraordinarily rich career, very inventive, very close to physics, very tremendous insights; and then, during the period on which he had to learn mathematics, particularly differential geometry in various forms, he changed. He changed his mind. That great originality for physics was altered ...*

Was Rabi right? Did Einstein change? The answer is, *Einstein did change.* Evidence for this change can be found in his Herbert Spencer lecture of 1933 bearing the title *On the Method of Theoretical Physics*:

> *... the axiomatic basis of theoretical physics cannot be extracted from experience but must be freely created ...*
>
> *Experience may suggest the appropriate mathematical concepts, but they most certainly cannot be deduced from it ...*
>
> *But the creative principle resides in mathematics. In a certain sense, therefore, I hold it true that pure thought can grasp reality, as the ancients dreamed.*

One may or may not agree with these very concise statements, but one has to agree that they powerfully and emphatically describe Einstein's perception in 1933 about how to do fundamental theoretical physics, a perception that represents a profound change from his earlier days.

Fig. 4. Einstein in Princeton.

So, Einstein did change. He himself was keenly aware of this change. In the *Autobiographical Notes*, published when he was 70 years old, we read:

> *... and it was not clear to me as a student that the approach to a more profound knowledge of the basic principles of physics is tied up with the most intricate mathematical methods. This dawned upon me only gradually after years of independent scientific work.*

It is evident that in this passage the *independent* scientific work was his long struggle to formulate general relativity during the period 1908–1915. That long struggle had changed him. Did he change for the better? Rabi would say: no, his changed perception had become a futile obsession. We would say: yes, his changed perception has altered the future course of development of fundamental physics.

Einstein's perception had permeated the very soul of the research in fundamental theoretical physics in the 50 years since Einstein's death, serving as a lasting testimonial to his courageous, independent, obstinate and perceptive greatness.

https://doi.org/10.1142/9789819826636_0012

FERMI'S β-DECAY THEORY

CHEN NING YANG
Tsinghua University, Beijing
and
Chinese University of Hong Kong, Hong Kong

Received 31 January 2012
Accepted 1 February 2012
Published 10 February 2012

Keywords: Enrico Fermi; beta-decay theory; history of physics.

I

Throughout his lifetime Enrico Fermi (1901–1954) had considered his 1934 β-decay theory as his most important contribution to theoretical physics. E. Segrè (1905–1989) had vividly written about an episode at the inception of that paper:[1]

> *Fermi gave the first account of this theory to several of his Roman friends while we were spending the Christmas vacation of 1933 in the Alps. It was in the evening after a full day of skiing; we were all sitting on one bed in a hotel room, and I could hardly keep still in that position, bruised as I was after several falls on icy snow. Fermi was fully aware of the importance of his accomplishment and said that he would be remembered for this paper, his best so far. He sent a letter to Nature advancing his theory, but the editor refused it because he thought it contained speculations that were too remote from physical reality; and instead the paper ("Tentative Theory of Beta Rays"*[a] *[FP 76]) was published in Italian and in the Zeitschrift für Physik.*[b]

In 2001 there was a centennial celebration of Fermi's 100th birthday. I contributed a paper to that celebration. One passage of my paper read:[3]

> *One day in the 1970's, I had the following conversation with Eugene Wigner in the cafeteria of Rockefeller University*:

[a]Original title: "Tentativo di una teoria dei raggi β."
[b]Fermi also published in 1933 a short paper in Italy. See Ref. 2.

Reproduced from International Journal of Modern Physics A, Vol. 27, No. 03n04, 1230005 (2012), DOI: 10.1142/S0217751X12300050

Y: *What do you think was Fermi's most important contribution to theoretical physics?*

W: *β-decay theory.*

Y: *How could that be? It is being replaced by more fundamental ideas. Of course it was a very important contribution which had sustained the whole field for some forty years: Fermi had characteristically swept what was unknowable at that time under the rug, and focused on what can be calculated. It was beautiful and agreed with experiment. But it was not permanent. In contrast the Fermi distribution is permanent.*

W: *No, no, you do not understand the impact it produced at the time. Von Neumann and I had been thinking about β-decay for a long time, as did everybody else. We simply did not know how to create an electron in a nucleus.*

Y: *Fermi knew how to do that by using a second quantized ψ?*

W: *Yes.*

Y: *But it was you and Jordan who had first invented the second quantized ψ.*

W: *Yes, yes. But we never dreamed that it could be used in real physics.*

What this passage reflected was not just the very different evaluations, by Wigner and me, of Fermi's β-decay theory, but in fact the very different evaluations of this theory by Fermi himself and his generation of physicists and by my generation. Recently I looked into the old literature and was able to understand better the reason for this difference.

II

In 1932 there were two discoveries that greatly shocked the world of physicists:

On February 17th, J. Chadwick (1891–1974) sent a short article to *Nature*[4] with the title "Possible Existence of a Neutron." It had great immediate impact: One realized that nuclei were made of protons and neutrons rather than protons and electrons, thus understanding the many regularities of the composition of light nuclei. But at the same time it created a new difficult puzzle: Since there were no electrons inside a nucleus, from where came the electron emitted in β-decay?

Seven months later on September 1st, C. D. Anderson (1905–1991) sent a short paper[5] to *Science* entitled "The Apparent Existence of Easily Deflectable Positives." Before this paper physicists had been dumbfounded for some five years by Dirac's famous equation of the electron: It was pure *magic*, but also afflicted[c] with a strange *sickness*. Should one believe it or not? Anderson's paper gave a definite resounding answer to this question: *Yes. One must believe it.* Furthermore Dirac's "hole theory," unbelievable as it was, was *nevertheless also correct.*

[c]Page 1056 of D. F. Moyer, *Am. J. Phys.* **49**, 1055 (1981) (Ref. 6).

After Anderson's paper, the remaining puzzling question was about β-decay. β-decay was first discovered near the end of 19th century. Through many efforts, finally in 1927, C. D. Ellis (1895–1980) and his collaborators confirmed that the electrons emitted in β-decay did not have a fixed energy. Instead they had a continuous spectrum. Niels Bohr (1885–1962) thought that this proved that in β-decay energy was not always conserved. But W. Pauli (1900–1958) proposed that energy was still conserved, but during β-decay, in addition to the electron, there was also emitted from the nucleus another particle which had no electric charge. (This particle was later named the neutrino.) But since at that time there were only three known particles: proton, electron and photon, Pauli did not dare to formally publish this bold proposal. For example, on December 1st, 1931 Pauli visited Princeton. A few days later[d] he had dinner in a Chinese restaurant in New York City with I. I. Rabi (1898–1988). Pauli told Rabi: "I think I am cleverer than Dirac, I shall not publish it." The "it" here referred to his neutrino hypothesis. Obviously Pauli thought Dirac was unwise to have published that year in the Proceedings of Royal Society an article about an undiscovered magnetic monopole.

Two years later, in October 1933 at the Solvay Congress in Brussels, Pauli was no longer so careful. He said at the conference:[7]

> *In June 1931, on the occasion of a conference at Pasadena, I proposed the following interpretation: The laws of conservation remain valid/recognized, the emission of the particles β being accompanied by a very penetrating radiation of neutral particles, which has not been observed until now. The total energy of particle β and of the neutral particle (or the neutral particles, since we do not know whether there is only one or there are a lot) emitted by the nucleus during a single process, would be equal to the energy which corresponds to the upper boundary of spectrum β. Needless to say we not only keep the conservation of energy, but also of momentum, of angular momentum and of the nature of the statistics in all the basic processes.*
>
> *Regarding the properties of these neutral particles, the atomic weights of the radioactive elements tell us first of all that their mass cannot exceed much that of the electron. In order to distinguish them from the heavy neutrons, Mr. Fermi proposed the name "neutrino." It's possible that the neutrinos' own mass might be equal to zero, so that they would have to propagate with the speed of light, like the photons. However, their penetrating power would exceed by far that of photons of the same energy. It seems to me acceptable that the neutrinos have spin $\frac{1}{2}$ and satisfy Fermi's statistics, although experiments do not give us any direct proof of this hypothesis. We know nothing about the interaction of the neutrinos with other material particles and with the photons: the hypothesis that they possess a magnetic moment, like I have proposed before (Dirac's theory predicted the*

[d]Page 1060 of D. F. Moyer, *Am. J. Phys.* **49**, 1055 (1981) (Ref. 6).

possibility of the existence of magnetic neutral particles), does not seem to me justifiable at all.

To clarify matters, the experimental study of the outgoing momenta in β-decay constitutes a problem of the most importance; we may predict that difficulties will be very great because of the small value of the energy of the recoil nucleus.

This paragraph was important: Every point in it turned out to be *correct*. Why did Pauli become so bold this time? We do not know the answer. But perhaps the discovery of the positron a year earlier, which had confirmed the unbelievable hole theory of Dirac, might have emboldened Pauli: After all, the hole theory was exceedingly bold and *it was based on Pauli's own Exclusion Principle*

III

The Solvay Congress in October 1933 was attended by many people: Bohr, Rutherford, Chadwick, Pauli, Heisenberg, Dirac and Fermi were all there. The comments by Pauli quoted above attracted much attention. But these comments did not answer a puzzling question: Nuclei are very small. To confine an electron inside a nucleus would cause its momentum to become very large by the uncertainty principle. It would then have a very large energy. How could that be?

Two months later, Fermi solved this puzzling problem. His key new ideas were:

(i) To add a term in the Second Quantized Hamiltonian

$$\Psi_e^\dagger(\cdots)\Psi_\nu\,. \tag{1}$$

(ii) Ψ_e, Ψ_ν are both anticommuting operators.

Second Quantized Hamiltonian is what is now called field theory. It was developed in the years 1925–1928 by Dirac, Jordan, Klein, Pauli and Heisenberg. Anticommuting operators were the inventions of Jordan and Wigner in 1928. Fermi was in fact at first not familiar with these operators.[8]

If so, why was it Fermi and not the architects of field theory who wrote this paper on β-decay that shocked the world of physicists? I think there are three reasons:

(a) In 1933 physicists thought matter was composed of protons, neutrons and electrons. In a chemical reaction or in a radiative decay such as in α or γ emission, protons, neutrons and electrons were all conserved. The only particle which was not always conserved was the photon. Thus it was difficult to understand the nonconservation of electrons in β-decay.

(b) From the pure theoretical point of view, field theory was built upon a *theorem* which stated that imposing the Bose–Einstein condition on a multiparticle Schrödinger equation is equivalent to a second quantized equation for field

Ψ_i which satisfies the commutation rule:

$$\Psi_i \Psi_j^\dagger - \Psi_j^\dagger \Psi_i = \delta_{ij} \, . \tag{2}$$

This fundamental theorem and its generalization to the Fermi–Dirac case required rather *complicated proofs*. These proofs were all confined to Hamiltonians for which the *electrons were conserved*.[9–11] It was thus generally believed that second quantization theory cannot be used for any problem with particle nonconservation. Fermi's bold proposal to add the term $\Psi_e^\dagger(\cdots)\Psi_\nu$ therefore shocked Wigner and von Neumann.

(c) Perhaps the most important reason was Fermi's unique approach to physics which was different from that of the architects of field theory: To him physics was not just formalism. Pure formalism to him was only useful if it materialized into explicit applications. A good example was his 1930–1932 paper on quantum electrodynamics,[12] in which he solved a fundamental formalistic problem in quantum electrodynamics and *then used it for calculations in five different applications.* This experience had *thoroughly acquainted* Fermi with how photons were created/annihilated by the Bose–Einstein operators $a^\dagger$ and a. I guess this experience had enabled him, two months after the Solvay Congress, to think of the idea of creating an electron by the Fermi–Dirac operator $b^\dagger$. In a paper by F. Rasetti published after Fermi's death, there is the following passage which supports this guess of mine:[13]

> *Apparently he had some difficulty with the Dirac–Jordan–Klein method of the second quantization of fields, but eventually also mastered that technique and considered a beta-decay theory as a good exercise on the use of creation and destruction operators.*

IV

Let us now return to the question posted in Sec. I above: Why was there such a big difference between the evaluations of Fermi's β-decay theory by two generations of physicists? I believe there are two reasons:

(A) By my generation, field theory was a required course for graduate theory students. Noncommuting Dynamic Variables was *in the first chapter* of field theory. We therefore did not realize that it was in fact a revolutionary development first launched by Fermi in that 1934 paper.

Historically each introduction of new numbers or operators was a revolution. Before the 16th century, people knew only of real numbers and considered many quadratic equations as unsolvable. In 1545 G. Cardano[14] published "Ars Magna" in which he introduced a new symbol in $5 + \sqrt{-15}$. But he was evidently not bold enough to further investigate the meaning of this symbol. It was twenty some years later that R. Bombelli in "L'Algebra" introduced for the first time the numbers i and $-i$ that we use today.

Quantum mechanics was launched by Heisenberg in 1925. With Born and Jordan he then proposed to replace dynamic variables with noncommuting operators. And that was of course a very very great revolution.

The physics of 19th century was entirely built on real numbers. Therefore when Schrödinger in 1926 wrote down his first wave equation, everything was real. It was only after a big struggle that Schrödinger was later forced to accept complex wave functions.[15]

(B) The most important development in experimental physics in the 1930's was the discovery in 1938 of nuclear fission. The most important development in theory in the 1930's was Dirac's hole theory. Compared with these two developments β-decay theory did not have the same degree of impact. Twenty some years later, with longer perspective, my generation thus failed to appreciate Fermi's impact as much as his contemporaries did in 1934.

V

P. Jordan had made two first class contributions to theoretical physics: Quantum mechanics was launched in three papers, the one-man paper, two-man paper and three-man paper. Of these Jordan was a co-author of the last two. Field theory was launched in three papers. The first was by Dirac in 1927, and the next two respectively by Jordan & Klein and Jordan & Wigner. The Jordan and Wigner paper was especially important because it proposed the very new anticommutation relations

$$b_i b_j^\dagger + b_j^\dagger b_i = \delta_{ij} \,. \tag{3}$$

Fermi's β-decay theory of 1934 was based on this relationship.

After World War II Jordan's name seemed to have disappeared from physics. I never saw any paper by him, nor did I encounter him at any scientific conference. His name was only mentioned in connection with his having been a member of the Nazi Party.

Between 1949–1966 I was at the Institute for Advanced Study in Princeton. Wigner was at that time a professor at Princeton University. As a consequence we saw each other almost weekly. I remember at teatime, if somebody mentioned Eq. (3) and said it was first written down by Wigner and Jordan, Wigner would always immediately say: No, no, it was Jordan and Wigner. After a few exchanges like this, even though everybody knew Wigner was super polite, we all felt that that paper was mainly the contribution of Jordan.

Wigner received his Nobel Prize in 1963. One day after that he mentioned in a casual conversation that Jordan had asked him to propose Jordan for the Nobel Prize. I remember Wigner told this story with an expression of helplessness. Thus nobody asked him any further questions. Apparently[16] Wigner did later propose Jordan for the 1979 Nobel Prize just before Jordan died in 1980.

I am indebted to C. F. Chen, Demetrios Christodoulou and B. F. Zhu for information and suggestions.

References

1. E. Segrè, *Enrico Fermi Physicist* (University of Chicago Press, 1970).
2. *The Collected Papers of Enrico Fermi* (University of Chicago Press, 1962), pp. 540–544.
3. C. N. Yang, in *Proc. Int. Conf. "Enrico Fermi and the Universe of Physics"*, Rome, September 29–October 2, 2001, eds. C. Bernardini *et al.* (ENEA – Ente per le Nuove tecnologie, L'Energia e l'Ambient, Roma, 2003), pp. 391–392.
4. J. Chadwick, *Nature* **129**, 312 (1932).
5. C. D. Anderson, *Science* **76**, 238 (1932).
6. D. F. Moyer, *Am. J. Phys.* **49**, 1055 (1981).
7. *Structure et Proprietes des Noyaux Atomiques: Rapports et Discussions*, Solvay Congress, October 1933, pp. 324–325 [translator: Weng Fan].
8. F. Rasetti, *The Collected Papers of Enrico Fermi*, Vol. 1 (University of Chicago Press, 1962), p. 539.
9. P. Jordan and O. Klein, *Z. Phys.* **45**, 751 (1927).
10. P. Jordan and E. Wigner, *Z. Phys.* **47**, 631 (1928).
11. W. Heisenberg, *The Physical Principles of the Quantum Theory* (Dover Publications, 1930), Sec. 11 of its appendix. [This book was the famous lecture notes of Heisenberg at the University of Chicago in the Spring of 1929. I had learned field theory from these notes.]
12. E. Fermi, *Rev. Mod. Phys.* **4**, 87 (1932).
13. F. Rasetti, *The Collected Papers of Enrico Fermi*, Vol. 1 (University of Chicago Press, 1962), p. 539.
14. B. L. Van Der Waerden, *A History of Algebra* (Springer-Verlag, 1985), pp. 56–61.
15. C. N. Yang, in *Centenary Celebration of a Polymath*, ed. C. W. Kilmister (Cambridge University Press, 1987), p. 53.
16. B. Schroer, Pascual Jordan, his contributions to quantum mechanics and his legacy in contemporary local quantum physics, arXiv:hep-th/0303241v2.

https://doi.org/10.1142/9789819826636_0013

C. S. Wu's Contributions: A Retrospective in 2015*

C. N. Yang
Tsinghua University, China

Published 20 July 2015

I

Chien-Shiung Wu started her research at Columbia University after World War II. The area she was working in, beta decay, was popular at that time. Wu had a solid foundation in theory and excellent ability to conduct experiments, so she quickly became famous. Around 1948, the major problem puzzling physicists working on beta decay was: Fermi's theory, Konopinski's theory, or other theories, which one is correct?

Many experiments gave different results, which was very confusing.

Wu realized that the thin films used in the majority of these experiments were too thick. She thus elaborately designed an extremely thin film to conduct a new experiment. After the experiment, she published a brilliant paper in 1949,[1] "*The Beta-Ray Spectra of* Cu^{64} *and the Ratio of* $N+/N-$", with the conclusion:

> *"The good agreement between the theoretical and experimental curves in Fig. 2 indicates that the Fermi theory probably does approximate the true distributions for negatrons and positrons at low energies. In any event, any remaining true deviations must be much smaller than has been previously suggested."*

Thus Fermi's theory was verified to be correct! And Wu became the leading physicist in the experimental field of beta decay!

*Based on a presentation dedicated to the 103rd anniversary of the birth of C. S. Wu, celebrated at the C. S. Wu Institute of Technology, China, 28 May 2015.

Reproduced from International Journal of Modern Physics A, Vol. 30, No. 20, 1530050 (2015), DOI: 10.1142/S0217751X15300501

II

In the same year, she published another paper,[2] "*The Angular Correlation of Scattered Annihilation Radiation*", studying the polarizations of the two γ in

$$e^+ + e^- \to \gamma + \gamma .$$

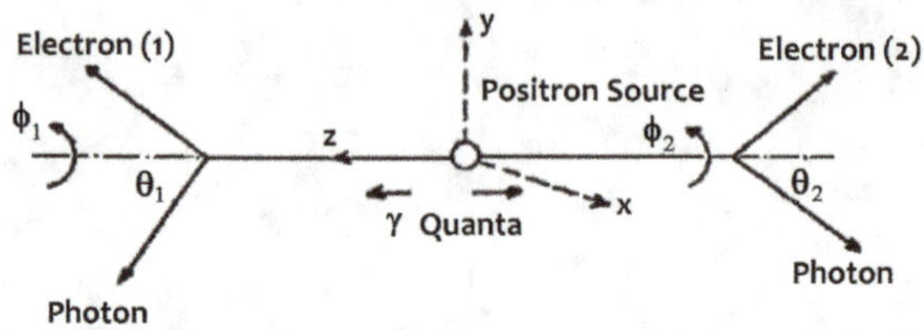

That was the first experiment on *quantum entanglement*, which is a very hot new area of research in the 21st century.

III

In the summer of 1956, Chien-Shiung Wu undertook the greatest experiment in her life.

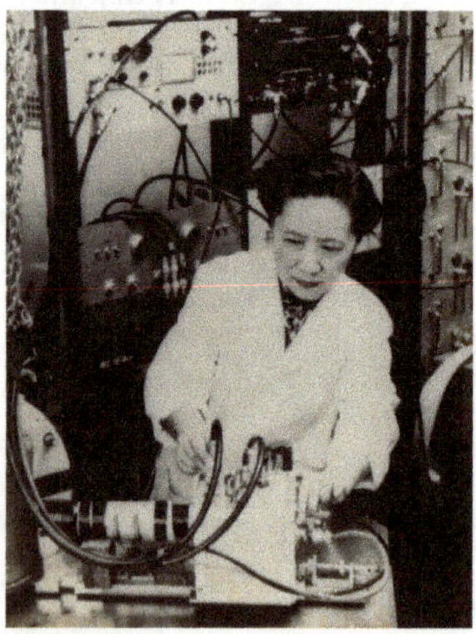

"Experimental Test of Parity Conservation in Beta Decay."[3]

She discovered that parity is not 100% conserved.

This discovery of parity nonconservation was front page news on *The New York Times*, January 16, 1957, under the headline

Basic Concept in Physics Is Reported Upset in Tests

Conservation of Parity Law in Nuclear Theory Challenged by Scientists at Columbia and Princeton Institute

IV

Eight years later, Wu published another very important paper,[4] "*Experimental Test of the Conserved Vector Current on the Beta Spectra of B^{12} and N^{12}*", with the following conclusion:

> *"This investigation confirms that the deviations from the allowed shape of the observed beta spectra for B^{12} and N^{12} have the correct magnitude and sign due to the weak magnetism term. This unique relation between the beta interaction and electrodynamics strongly supports the conserved vector-current theory."*

Thus she verified the conservation of vector current (CVC), thereby influencing later theories:

- The Electroweak Theory
- The Standard Model

In a comment in April 1997, I wrote:

> *"C. S. Wu's work is world-renowned for her accuracy, but her success derives from an even more important factor: In 1956, nobody was willing to do an experiment to test parity conservation, why then was she willing to undertake this difficult task? Because she alone had this wisdom, believing that even if parity was not overthrown, this fundamental law should nevertheless be proven experimentally. This is where she surpassed all others."*

References

1. C. S. Wu and R. D. Albert, *Phys. Rev.* **75**, 1107 (1949).
2. C. S. Wu and I. Shaknov, *Phys. Rev.* **77**, 136 (1950).
3. C. S. Wu, E. Ambler, R. W. Hayward, D. D. Hoppes and R. P. Hudson, *Phys. Rev.* **105**, 1413 (1957).
4. Y. K. Lee, L. W. Mo and C. S. Wu, *Phys. Rev. Lett.* **10**, 253 (1963).

https://doi.org/10.1142/9789819826636_0014

C. N. Yang on Einstein and Newton*
October 2015

Published 27 November 2015

In Professor C. N. Yang's view, Einstein's strength was in his ability to distinguish what was truly important and to investigate it. Also, Einstein was unique in that he was able to zoom in as well as zoom out, just like a film which has both close-up and long shots. Many people are only able to have one view, either close-up or from afar, and cannot switch between the two. Professor C. N. Yang feels that, in the history of physics, only Newton can be compared with Einstein. Although Maxwell and Boltzmann were prominent physicists, their influence was not as great as Einstein's.

Of the three greatest contributions to physics in the 20th century, two and a half came from Einstein — this is Professor Chen-Ning Yang's evaluation of Einstein. As one of the greatest scientists of the last 1000 years, Einstein surely deserves such an evaluation. The "three greatest contributions to physics in the 20th century" that Professor C. N. Yang refers to are special relativity, general relativity and quantum mechanics. Taking relativity as an example, the author of an article in this issue checked a preprint library and discovered that there were more than 2000 papers on relativity and related topics in 2014 alone. Furthermore, many theories such as the quantum Hall effect and the multiverse hypothesis, etc. were all developed out of quantum mechanics and the theory of relativity and stand on the frontiers of modern research in physics. This shows how much the theory of relativity has influenced the field of physics in the past 100 years.

On several occasions related to the 125th anniversary of the birth of Einstein, Professor C. N. Yang gave a number of lectures on Einstein. Therefore, to mark the 60th anniversary of Einstein's death and the centenary of the publication of the general theory of relativity, we have specially invited Professor Yang and his old friend, Jiang Caijian (author of a biography of C. N. Yang and a former science

Reproduced from International Journal of Modern Physics A, Vol. 30, No. 33, 1530064 (2015), DOI: 10.1142/S0217751X15300641

correspondent for *China Times*, Taiwan), to discuss Einstein's achievements, his unique attributes and the future of physics. The interview was conducted in Professor Yang's office at the Institute for Advanced Study, Tsinghua University.

Jiang: *We have specially arranged this interview because 'Scientific American' is publishing a special issue to mark this year, the 60th anniversary of Einstein's death as well as the centenary of the publication of the general theory of relativity. This is also the 136th anniversary of Einstein's birth. Eleven years ago, on March 14, 2004, you were invited to give a lecture at a conference in Germany (held at Einstein's birthplace, Ulm) to mark the 125th anniversary of Einstein's birth. The title of your lecture was "Einstein's Impact on Theoretical Physics of the 21st Century". Let us start by discussing that very profound lecture.*

You began by referring to Einstein's special theory of relativity and spoke on two very interesting issues: that the term "relativity" was not coined by Einstein but by Poincaré; and that the transformation formulas of special relativity were first written down not by Einstein, but by Lorentz, after whom the transformation is still named. You said that Poincaré understood the philosophy of relativity but not its physics, and that Lorentz knew the mathematics of relativity but not its physics. Only the young Einstein could create this revolution.

You said that you had particular admiration for Einstein from your early days as a student. What do you think is so special about Einstein?

C. N. Yang: Einstein's brilliance was, on the one hand, in mathematics. He could intuitively appreciate the subtleties of mathematics. On the other hand, he also understood physical phenomena up close. He was different from others in that he could zoom in and zoom out, like in a film with both close-up and long shots. This ability to switch between different views is powerful. Many people are only able to have one view, either close-up or from afar, and cannot switch between the two.

Jiang: *This is what you said to Huang Kun at Southwestern Associated University: look from above or from below.*

C. N. Yang: Yes. We research students at the Southwestern Associated University (SAU) had already reached that level of understanding of physics. We recognized the benefits of being able to see the scene from on high, where one can switch between views, and therefore see more than is possible from the single viewpoint of one who looks only from the ground.

Jiang: *Since we are talking about SAU, can you say a little more about it? The university was established in a very difficult period, during the war of resistance against Japan. The university lasted only nine years, and had little in the way of funds or international connections; yet it achieved the greatest success in the history of universities in China. Given the current trends in education (such as a strong emphasis on internationalization), the example of SAU does make one think. As an outstanding representative of that environment, please say a little more about the experience.*

C. N. Yang: This is my view. Many people have discussed the reasons for SAU's success. The reasons are many, not just one. One reason neglected by many, but which is in fact very important, is that SAU happened to have been established at a crucial time, so that graduates of the university were later able to participate in the construction of a new China and in the economic take-off in Taiwan. Those of my age and those about ten years younger found themselves in a time of opportunity, when China was emerging from the position of a closed agricultural country to develop into an open industrialized country; Taiwan went through a similar economic path. As a result, my generation and those about ten years younger had good opportunities for achievements, which were closely tied to the opportunities of that time. The importance of this point is much underrated by many.

Jiang: *Let us return to relativity. The theory of relativity led to the concept of symmetry and became a cornerstone of 20th century physics research. Einstein's special relativity arose from electromagnetic theory and led to an understanding of symmetries. Later, the reverse was true for his theory of general relativity: it was derived from concepts of symmetries. You have said that he had both great foresight and great courage. Can you explain a little about Einstein's creativity?*

C. N. Yang: In any area of cutting-edge scientific research, there are many unclear issues. The eternal question is which of these are worth pursuing, and which are a waste of time to investigate. Einstein was especially proficient at distinguishing between the two. He spent eight years investigating symmetries, because he recognized their importance. Another example: in 1923 or 1924, the young Indian Satyendra Nath Bose mailed him an English manuscript which had been refused by English journals and asked Einstein to translate it into German for publication in Germany. This impolite request did not annoy Einstein. He not only translated the manuscript into German and published it under Bose's name, but also added a short remark saying that the paper made much sense and he himself would wish to continue to develop it. After that, he published several papers to develop an idea later called Bose–Einstein condensation. This theory was so astonishingly revolutionary that his contemporaries thought he was insane. Even his friend, Dutchman Paul Ehrenfest, opposed him. However, Bose–Einstein condensation was eventually recognized by the physics world as an important and correct theory in the 1950s. This story illustrates how Einstein was able to instantly recognize a deep point which Bose was unable to see. He was able to further explore it and produce a revolutionary contribution.

Speaking of Ehrenfest, he was a friend of Einstein's who later committed suicide in 1933. After his death, Einstein wrote a paper in which he praised Ehrenfest's creativity. He went on to say that Ehrenfest's critical faculty was greater than his creativity, which was why many of his creative ideas were stillborn.

Jiang: *After Einstein developed the theory of general relativity, he also thought of three ways to test the theory. Was this his unique style?*

C. N. Yang: Yes. This was very important, and it is something that is often insufficiently valued when discussing Einstein. After Einstein wrote down his equations, he immediately turned to the question whether there was any way to test these equations. This involves both physics and mathematics. The papers that he produced in the next year or two proposed ways to experimentally test the general theory of relativity. These experiments did brilliantly confirm his theory and his reputation grew until he became a household name throughout the world. That is why, if we talk about great physicists, I feel that only Newton can be compared with Einstein. Maxwell and Boltzmann were also great physicists, but their influence was not as great as Einstein's. Of the three greatest contributions to physics in the 20th century, two and a half were due to Einstein.

Jiang: *Let's talk a little more about Einstein. He said that theories of physics can be "derived" from mathematics. You said earlier that Einstein was able to switch between theory and experiments. Is there any contradiction between these two statements?*

C. N. Yang: No, because Einstein's views of the value of mathematics varied at different times. In one of his later papers, he said that when he was younger, he had not appreciated the function of mathematics in physics. In those days, he had approached physics in a more intuitive way and had an intuitive interest in physical phenomena. It was only after developing his general theory of relativity that he appreciated the importance of mathematics in physics. In the 1930s, he went on to say that creative thinking in theoretical physics came out of mathematics, a view he held until his death.

Jiang: *You have mentioned Einstein's intuition and feelings towards mathematics and physics. Your good friend Freeman Dyson said his own intuition in mathematics was stronger. In your work on the Yang–Mills gauge field theory in 1954, did it arise out of physics or mathematics?*

C. N. Yang: That was the most important paper I have ever written. Recently, Dyson said that it was the most important paper since Dirac's 1928 paper. We wrote the paper because the time was right on two counts. First, after World War II, scientists discovered many new particles and there was much research into their interactions. Many papers were published proposing different interactions. I felt that this was unsatisfactory and that there had to be a general principle governing the interactions between particles. Second, at the time, I much admired H. Weyl's gauge invariant theory of 1929 and wished to try to broaden it to find a general principle. These two reasons for writing the 1954 paper arose from both physics and mathematics.

Jiang: *The 1954 Yang–Mills gauge field theory and the development of symmetries had a great impact on 20th century physics. Yang–Mills theory and Einstein's general theory of relativity have similarities in their equations — both highly non-linear. It seems you did not realize this until many years later.*

C. N. Yang: You are right. It was only in the 1960s when I was at Stony Brook (New York State University at Stony Brook) teaching general relativity that I realized this. After that, I consulted my friend mathematician Simons, and in 1975, Wu Tai Tsun and I fully clarified an important concept: the mathematics of gauge field theory was exactly the fiber bundle theory in mathematics.

Jiang: *In* **The Future of Physics — Revisited** *you recalled that at the 100th anniversary of MIT in 1961, you, Feynman and two others were invited to participate in a discussion forum. At that time, you were 39 years old and the youngest of the four. However, according to the transcript, your views were already very profound and not very optimistic. In contrast, Feynman was very optimistic. He said in closing, "Fundamental physics has a finite lifetime (because the ultimate theories would have been grasped). It has (only) a while to go. I take advantage of the fact that I live in the right age, but I do not think it will go on for a thousand years."*

Did you notice this?

C. N. Yang: Yes, I did. Feynman had several unusual intuitive views about physics. His path integrals and Feynman diagrams were great intuitive contributions of the century. I guess that around 1961, he probably had new intuitive inspiration, which was why he was so optimistic. What the intuitive inspiration was is worth investigating. I think our cultural background also contributed to our differences in views. I had been trained in traditional Confucianism to "continually examine my actions". He was a product of America.

On this topic, we naturally would ask, "American scientists are often proud and arrogant, and American science is very successful. Are these two observations related?" In other words, if young people in China were to imbibe this American trait, would it enable them to be more successful in science? This is worth discussing.

Reporter from Global Science (Chinese): *Prof. Yang, although you have said that this is worth discussing, I believe that your opinion is that there is a relationship between American scientists' manner and their success. Is that right?*

C. N. Yang: You are right. I do believe that American scientists' pride and their success are inter-related. In extremely competitive activities, gentlemen often lose out.

Jiang: *This is exactly what I am going to ask next.*

C. N. Yang: From the perspective of non-scientists, the attitudes towards science expressed in the speeches by Feynman and myself in 1961 were in clear contrast. Wasn't that also a reflection of our different cultural traditions? Which attitude towards scientific development was better? I think these are very interesting questions.

Jiang: *I have not had the opportunity to meet Feynman, but I think he must have had good stage presence and eloquence. But looking at his papers, he did not seem to have much depth, using his reasoning method to explore phenomena. I think there is a question here. Although Feynman was obviously talented, he was also daring. His paper on path integrals incorporated three important themes you mentioned in a paper: quantum, symmetries and phase. Modern science reflects western views about the universe. You also engaged in modern scientific research, but your cultural background was different. From this angle, I feel that culture influences a person's view about the universe and about the world. Aside from the attitude towards science, does this also influence a person's attitude about everything else?*

C. N. Yang: Of course it has an influence. Feynman's 1961 paper should be compared to James Watson's 1968 book *The Double Helix* (Watson discovered the double helix in DNA). Both exhibited the American spirit. This is "you do not need a lot of knowledge or great past accomplishments; as long as you have a fearless spirit and are willing to strive hard, you will succeed". I must add that Feynman had deep intuitive understanding of some aspects of physics. I feel that his path integral is one of the great historic contributions to physics. However, he did not appreciate traditional theoretical research methods, and that is a shortcoming.

Jiang: *Watson's book vividly describes this American spirit, and from the sales point of view, or from the view of describing the intense competition in scientific research, it has been successful. However, traditional gentlemen will not like the way the book is written.*

C. N. Yang: Cultural background does influence the attitude toward scholarship. Theoretical physics is based on conjectures. These conjectures are not based on logical deduction and may be either intelligent or stupid. Like playing chess, an expert may not be able to tell you why he makes a particular move, but his conjecture will be correct. In making conjectures, some are more daring while others are less so. For a young person, should we encourage him to be more daring or less daring? I think Chinese culture would advise him to be less daring, while American culture would advise him to be more daring.

Jiang: *Back to your 1961 paper, you said that the mysterious structure of the universe was infinite while human intelligence was finite, and it would never be possible to use finite intelligence to fully grasp the infinite universe. In this respect, you and Einstein have similar views. Einstein said that he only knew of two things that were infinite: one was the universe, and the other was human ignorance, and he was not yet able to confirm the former. We know that your friend Dyson in his Einstein lecture* **Birds and Frogs** *said that some scientists were like birds, with a bird's-eye view, and he said that you were one of those.*

We can see that many new theories today are deduced from mathematics, but there is a question whether they are relevant to physics. This is an ultimate philosophical question and also a scientific question.

C. N. Yang: True. For instance, superstring theory has strongly influenced mathematics, but it is still unknown whether it has any relation to physical phenomena. I am also not so optimistic about supersymmetry, for two reasons. First, the mathematics used by fundamental physics is generally beautiful subtle mathematics. There are many examples of this. However, supersymmetry does not use subtle mathematics, so I surmise that it cannot be a cornerstone of fundamental physics. I had discussed this at a forum in 1979 to mark the 100th anniversary of Einstein's birth. Also, after several decades of work, there have been no experimental results confirming supersymmetry. So I am sceptical about its future.

Jiang: Scientific American's *special issue includes an article about Einstein's mistakes. Do you think that Einstein was wrong in holding these views?*

C. N. Yang: The first shot in cosmology was fired by Einstein. This was one of his great contributions. He could not have asked the questions we ask today. Whether or not Einstein introduced some parameters in his cosmology, we cannot use the knowledge we have today to evaluate Einstein's views.

Also, astronomy has developed very quickly in recent years. I think it might give rise to a great revolution in physics.

Acknowledgments

Special thanks to Professor Kenneth Young (Chinese University of Hong Kong) and Professor C. H. Oh (National University of Singapore) for reading this manuscript.

*This interview is published in the *International Journal of Modern Physics A* with permission by Prof. C. N. Yang. The Chinese version of the interview appears in《环球科学》(Global Science, China) Issue 10, 2015, pp. 24–27.

https://doi.org/10.1142/9789819826636_0015

Shoucheng Zhang — A physicist of the first rank*

C. N. Yang
Institute for Advanced Study, Tsinghua University, P. R. China

Published 12 February 2019

In condensed matter physics, researchers investigate properties of different kinds of matter: why copper can conduct electricity, while rubber cannot; why water can freeze into ice, but also evaporate into steam; etc. This area of physics has deep relations with applications, with the economic development of the world, and with daily lives of mankind. Thus, it is an especially important area of research. S. C. Zhang has led several revolutionary advances in this area of physics, of which the most important is in Quantum Spin Hall Effect and related areas.

In 2005–2006, C. L. Kane of the University of Pennsylvania and Zhang independently published theories pointing to the possible existence, in composite matter under proper conditions, of surface electric conductivity, an entirely new phenomenon. These two papers immediately caught the attention of all condensed matter physicists. But which composite matter, under what conditions, would exhibit this remarkable new phenomenon remained a very difficult problem.

Zhang told me that together with several semiconductor collaborators, he next made calculations for specially designed quantum wells in composite semiconductors, and published on 15 December 2006, a paper predicting that in a Hg–Te–Cd semiconductor quantum well there could be surface electric conductivity plus other new important phenomena. Now, there are many types of semiconductors, and quantum wells can have many different shapes and dimensions, how did they choose to focus on their specific Hg–Te–Cd quantum well? I think the only answer is: Zhang had deep intuition in the quantum mechanics of semiconductors.

In 2007, Molenkamp of Würzburg University and his experimental group, following Zhang's prediction, tried several kinds of HgTe/(HgCd)Te quantum wells, and created the first example of surface electric conductivity and related phenomena. In November that year, they published, together with Zhang and his student

*An earlier version of this paper, in Chinese, had appeared in GuangMing Daily-net in December 2018.

Reproduced from International Journal of Modern Physics B, Vol. 33, No. 04, 1930001 (2019), DOI: 10.1142/S0217979219300019

X. L. Qi, a paper announcing this great success. This paper is one of the most important papers in physics in recent years.

It has been the general opinion of all physicists that Kane, Zhang and Molenkamp will receive the Nobel prize in time. Now that Zhang has passed away, I firmly believe Kane and Molenkamp will one day be awarded the Nobel prize.

Reflections on Scientific Questions

https://doi.org/10.1142/9789819826636_0016

BOSE–EINSTEIN CONDENSATION IN A TRAP*

CHEN NING YANG

†*Institute for Theoretical Physics, State University of New York, Stony Brook, New York 11794, USA*

‡*Physics Department, The Chinese University of Hong Kong, Hong Kong*

Received 15 August 1996

In an amazing series of technological breakthroughs Bose–Einstein Condensation (BEC) of weakly interacting atoms was finally experimentally achieved in 1995, seventy-one years after the equally amazing original theoretical predication of Albert Einstein. The atoms were cooled to very low temperatures, (of the order of 10^{-8} K or less), in ingenious traps, in sufficient quantities for BEC to take place.

From the theoretical point of view, an interacting system of atoms is very difficult to investigate rigorously. But for atoms at extremely low temperatures and low densities, expansion in a parameter ρa^3, where ρ is the density of atoms and a the scattering length, was developed in the 1950 by three different methods: pseudopotential method,[1] binary collision method[2] and Green's function method.[3] Of these the pseudopotential method is particularly adaptable to the present experimental arrangements. But to add a trap to the system requires a further approximation, called the local density approximation (LDA), which is in the nature of the Thomas–Fermi method well known in atomic structure. In two recent papers[4,5] this approach was pursued for the case of positive scattering length ($a > 0$), resulting in approximate expressions for the density distribution $\rho(\mathbf{r})d\mathbf{r}$ and the double density distribution $\rho(\mathbf{r},\mathbf{p})d\mathbf{r}d\mathbf{p}$ of the atoms. I shall not go into the details of these calculations, but would like to make a few remarks:

(a) It is tempting to approximate the influence of the close-by atoms by an added potential V_a on each atom. While this idea is physically quite appealing, it must be used with care: For the gaseous phase the expression for V_a is $V_a = 8\pi a\rho\hbar^2 m^{-1}$ while for the condensed phase it is $V_a = 4\pi a\rho\hbar^2 m^{-1}$.

(b) For the case of no traps, the investigations of Refs. 1, 2 and 3 had been carried out in the 1950's to high orders of a. For example, for the ground state

*Contribution to the Proceedings of the Inauguration Conference of Asia Pacific Center for Theoretical Physics, 4–10 June, 1996, Seoul, Korea.

†E-mail: chen.yang@sunysb.edu

‡E-mail: cnyang@cuhk.edu.hk

Reproduced from International Journal of Modern Physics B, Vol. 11, No. 06, pp. 683–684 (1997), DOI: 10.1142/S0217979297000381

energy of a collection of hard spheres of diameter a, the result[6] is

$$E = \frac{N\hbar^2}{2m} 4\pi a\rho \left[1 + \frac{128}{15\sqrt{\pi}} \sqrt{\rho a^3} + 8 \left(\frac{4\pi}{3} - \sqrt{3} \right) \rho a^3 \ln(\rho a^3) + \cdots \right],$$

where N = number of particles. If we focus attention only on the first term (i.e. to order a^1), then the pseudopotential provides the easiest approach. To this order the pseudopotential is equivalent[7] to the δ-function interaction:

$$V_\delta = 2\pi a\hbar^2 m^{-1} \int d\mathbf{r} \psi^\dagger(\mathbf{r}) \psi^\dagger(\mathbf{r}) \psi(\mathbf{r}) \psi(\mathbf{r}) . \tag{1}$$

For the case with traps, to order a^1, (1) provides the easiest approach. Higher orders in a have not yet been studied, to my knowledge.

(c) What has been experimentally produced is a small droplet of superfluid in the center of the trap. The droplet should exhibit two fluid behavior which had been theoretically studied[8] from first principles.

References

1. Kerson Huang and C. N. Yang, *Phys. Rev.* **105**, 776 (1957); T. D. Lee, Kerson Huang and C. N. Yang, *Phys. Rev.* **106**, 1135 (1957). Also Ref. 8 below.
2. T. D. Lee and C. N. Yang, *Phys. Rev.* **105**, 1119 (1957); **113**, 1165 (1959); **116**, 25 (1959); **117**, 12, 22, 897 (1960).
3. S. T. Beliaev, *Sov. Phys. JETP.* **7**, 289, 299 (1958) and many later papers.
4. T. T. Chou, Chen Ning Yang and L. H. Yu, *Phys. Rev.* **A53**, 4257 (1996).
5. T. T. Chou, Chen Ning Yang and L. H. Yu, to be published.
6. The first two terms were first obtained in Refs. 1 and 2 above. The last term was obtained in T. T. Wu, *Phys. Rev.* **115**, 1390 (1959); N. Hugenholtz and D. Pines, *Phys. Rev.* **116**, 489 (1959) and K. Sawada, *Phys. Rev.* **116**, 1344 (1959).
7. This formula must be used with care: If one includes all orders of a, then for $a > 0$, (1) is equivalent to zero and for $a < 0$, (1) is undefined.
8. T. D. Lee and C. N. Yang, *Phys. Rev.* **112**, 1419; **113**, 1406 (1959). See also C. N. Yang, *Physica* **26**, S49 (1960).

https://doi.org/10.1142/9789819826636_0017

NECESSARY SUBTLETY AND UNNECESSARY SUBTLETY*

C. N. YANG
C. N. Yang Institute for Theoretical Physics, State University of New York, Stony Brook, NY 11794-3840, USA
yang@insti.physics.sunysb.edu

Received 2 November 2002

We are all very happy that Ray is receiving the honor that he so richly deserves.

Einstein had said "Subtle is the Lord ...".

Indeed nature presents layers upon layers of subtlety. To us at every moment, some of these subtleties appear necessary, others not.

Take as an example, quantum theory. According to classical physics around the beginning of the 20th century, all matter should be unstable. Some new concepts were urgently needed to stabilize matter. Thus quantum theory was born, as a *necessary* subtlety at the time. Or take string theory now. It was invented to amalgamate quantum theory with gravity. Should it be successful, it would be a *necessary* subtlety.

Now consider strange particles around the late 1940s. It was uncalled for. Their appearance was therefore an *unnecessary* subtlety. Or parity nonconservation in the mid-1950s. Again uncalled for. Therefore another *unnecessary* subtlety. It remains so to this day.

An *unnecessary* subtlety may in time become a *necessary* subtlety. Consider, for example the theory of energetics. To its adherents such as Mach and Ostwald, kinetic theory, if relevant, was an *unnecessary* subtlety. Of course, in time, kinetic theory became a *necessary* subtlety.

Personally I tend to disbelieve *unnecessary* subtleties. That is perhaps why Dyson called me conservative. I did not believe in neutrino oscillations, even after Davis' painstaking work and Bahcall's careful analysis: The oscillations were, I believed, uncalled for. Now, after the beautiful experiments which we shall hear about in the next few days, I have to surrender and accept neutrino oscillations as reality, but still as *unnecessary* subtleties.

*Speech at *Neutrinos and Implications for Physics Beyond the Standard Model*, Stony Brook, October 11–13, 2002.

Reproduced from Modern Physics Letters A, Vol. 17, No. 34, pp. 2229–2230 (2002), DOI: 10.1142/S0217732302008903

Like all physicists, I have faith in our discipline, and also believe that in time, neutrino oscillations will become, through new depths of understanding, *necessary* subtleties.

https://doi.org/10.1142/9789819826636_0018

EVOLUTION OF THE CONCEPT OF THE VECTOR POTENTIAL IN THE DESCRIPTION OF FUNDAMENTAL INTERACTIONS

A. C. T. WU

Harrison Randall Laboratory of Physics and Michigan Center for Theoretical Physics, University of Michigan, Ann Arbor, MI 48109-1120, USA
actwu@umich.edu

CHEN NING YANG

Chinese University, Hong Kong, China
and
Tsinghua University, Beijing 100084, China

Received 3 May 2006

Modern theories of fundamental interactions are based on the concept of "connections," a generalization of the vector potential **A**. How did this concept evolve in physics? We attempt in the present paper to trace this evolution, starting from the early 19th century. A schematic diagram which charts this evolution is exhibited (Fig. 1).

Keywords: Vector potential; Faraday; Maxwell; $(\mathbf{p}-e\mathbf{A})$; action principle; gauge theory; symmetry dictates interaction.

Contents

Reproduced from International Journal of Modern Physics A, Vol. 21, No. 16, pp. 3235–3277 (2006), DOI: 10.1142/S0217751X06033143

1. Introduction

Profound changes took place in the 20th century in physicists' understanding of the concepts of *space, time, motion, energy* and *force*. These are primordial but essential concepts important for the very survival of early man. They must therefore have entered, in one form or another, into the early vocabulary of every primitive human civilization. That these elementary but essential concepts should now acquire such precise and sophisticated mathematical meaning in physics will probably forever remain one of the great mysteries of nature, and of the epistemological history of mankind.

The profound changes over our understanding of *force*, or *interaction* in today's terminology in physics, took place gradually. It had started with the idea of action at a distance. Then in the mid-19th century, through the works of M. Faraday and J. C. Maxwell, one type of force, the electromagnetic force, became described by a *field theory*. It is no exaggeration to say that field theory of electromagnetism has proved enormously successful in the 20th century. But what about the other types of forces discovered in the 20th century, such as nuclear forces and the weak forces? What kind of field theory do they obey? It turns out that these forces are also

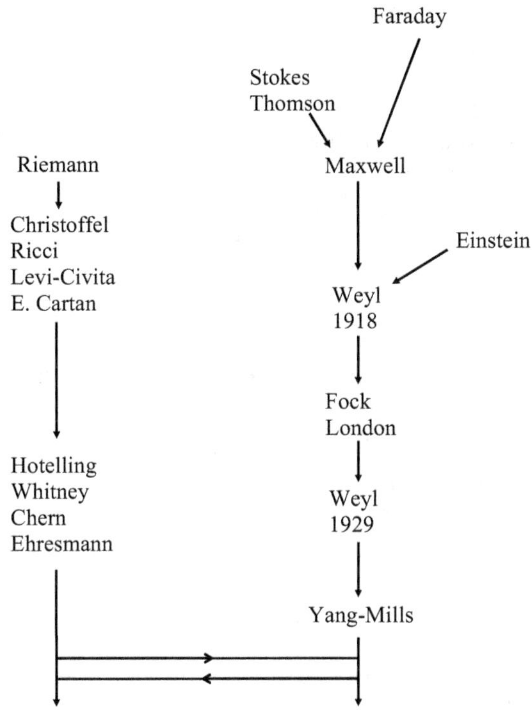

Fig. 1. Flow of ideas in the evolution of the concept of the vector potential.

describable beautifully and precisely by field theories, and that all these theories have mathematical structures required by the concept of symmetry. Hence the principle: *symmetry dictates interaction.* The conceptual history of this remarkable development is the subject of the present paper.

Playing an important part in this history is the vector potential **A**, which first made its appearance in the 19th century. There was certain freedom, now called gauge freedom in its definition, which was early recognized as a simple but somewhat annoying mathematical property. It is this freedom which has now metamorphosed into the key symmetry principle that dictates the exact equations describing the fundamental forces of nature.

Very remarkably, the mathematics of this symmetry principle was in the meantime developed by geometers in the theory of *fiber bundles*, entirely independently of the developments in physics. When this became known, a renewed cross-fertilization of basic ideas between the disciplines of physics and mathematics happily resulted.

Throughout this paper our emphasis is on the early motivation and evolution of the key ideas. There is a vast literature about various aspects of the history we try to cover. We shall thus not be able to include many important developments. On the other hand, we cite many references for the convenience of the readers, even though we do not necessarily agree with the views expressed in some of them.

2. Early Work on the Vector Potentials: Neumann, Weber, Kirchhoff and Thomson

2.1. *On the European continent: Franz Ernst Neumann (1798–1895); Wilhelm Eduard Weber (1804–90); Gustav Robert Kirchhoff (1824–87)*

The vector potential was first introduced to describe the magnetic field produced by an electric loop or a circuit element. E. T. Whittaker[132] credits F. E. Neumann[91] (in 1845) as the one who, in discussing the induced electrical currents from Ampère's law,[2] extracted from the interaction energy (between two current elements ds and ds' carrying current i and i') the expression for a vector potential

$$\mathbf{a} = \int \frac{i'\, d\mathbf{s}'}{r} . \tag{2.1}$$

Whittaker (Ref. 132, p. 230) attributed another expression to W. E. Weber (1846):[124]

$$\mathbf{a}' = \int \frac{i'(\mathbf{r}\cdot d\mathbf{s}')\mathbf{r}}{r^3} . \tag{2.2}$$

The difference $\mathbf{a} - \mathbf{a}'$ is a gradient of a scalar on account of an identity attributed to H. von Helmholtz (1870):[43]

$$\{(d\mathbf{s}\cdot d\mathbf{s}') - (\mathbf{r}\cdot d\mathbf{s})(\mathbf{r}\cdot d\mathbf{s}')/r^2\} = (d^2 r/ds\, ds')ds\, ds' \tag{2.3}$$

This is seen by noting that

$$\frac{\partial r}{\partial s} = \cos\theta\,, \quad \frac{\partial r}{\partial s'} = \cos\theta'\,,$$

where θ, θ' are the angles between $(\mathbf{r}, d\mathbf{s})$, $(\mathbf{r}, d\mathbf{s}')$ respectively. Equation (2.3) is the result by performing the second derivative $\partial^2 r/\partial s \partial s'$. Alternatively, a direct calculation gives

$$\mathbf{a} - \mathbf{a}' = \int i'\, d\mathbf{s}' \cdot \nabla\left(\frac{\mathbf{r}}{r}\right) . \tag{2.4}$$

It should be noted that Ampère[2] had worked out expressions very similar to (2.3). Maxwell gave an account of Ampère's theory in his *Treatise*[88] (see Ref. 88, Vol. 2, articles 502–527). The right-hand side would vanish when integrated over a closed loop. By hindsight, this is a historically interesting first example of gauge freedom.

In his work on the propagation of the electrical disturbance in the telegraph wire and later for a general conductor, Kirchhoff (1857)[65,66] had worked with Weber's vector potential (2.2).

2.2. *William Thomson (Lord Kelvin) (1824–1907)*

Among W. Thomson's many contributions, two of his papers[113,114] are of particular interest to us.

(I) In a 1847 paper,[113] "On a Mechanical Representation of Electric, Magnetic and Galvanic Forces," Thomson considered the cases when the expression (in modified notation, $\mathbf{u} = (\alpha, \beta, \gamma)$, $d\mathbf{r} = (dx, dy, dz)$)

$$\mathbf{u} \cdot d\mathbf{r} = df = \text{a complete differential}\,. \tag{2.5}$$

The three examples discussed by Thomson are:

(i) Electric case: $\mathbf{u} = \mathbf{E}$ field of unit point charge $= \mathbf{r}/r^3$; $-f =$ the corresponding electrostatic potential $= 1/r$.
(ii) Magnetic case: $\mathbf{u} = \mathbf{B}$ field of unit dipole with the unit normal $\mathbf{n}$ parallel to the dipole moment;

$$\mathbf{B} = \operatorname{curl} \mathbf{A}\,, \quad \mathbf{A} = \mathbf{n} \times \mathbf{r}/r^3\,, \quad f = \mathbf{n} \cdot \mathbf{r}/r^3\,.$$

(iii) Galvanic current case: $\mathbf{n}$ is parallel to the current, $\mathbf{u}$ in Eq. (2.5) is replaced by the Laplacian $\nabla^2\mathbf{u}$, $\mathbf{u} = \frac{1}{2}\nabla(\mathbf{n} \cdot \mathbf{r}/r) - \mathbf{n}/r$, $f = -\mathbf{n} \cdot \mathbf{r}/r^3$. Here $\operatorname{curl} \mathbf{u} = \mathbf{A}$ of (ii), and $\operatorname{div} \mathbf{u} = 0$.

Thomson cited the influence of Stokes' 1845 work on the equation of equilibrium of an elastic solid, and the statement (2.5) is related to the idea behind the so-called "Stokes' Theorem" (see the Appendix).

(II) In an extensive 1851 study, "A Mathematical Theory of Magnetism,"[114] Thomson, while working with quantities associated with magnetization distributions, had written down the solenoidal condition for a generic vector, say $\mathbf{B}$,

$$\operatorname{div} \mathbf{B} = 0\,, \quad \mathbf{B} = \operatorname{curl} \mathbf{A}\,. \tag{2.6}$$

Thomson clearly recognized that such vector $\mathbf{A}$ is to some extent arbitrary and is determined *modulo a gradient of a scalar.*

It is interesting to note that Thomson in 1850 while working on magnetism had discovered the integral identity which has came to be known as the Stokes' theorem. For a brief historical note, see the Appendix. In his letter to Stokes dated July 2, 1850 in which he communicated the integral theorem to Stokes, Thomson cited the influence of Stokes' papers;[108,109] in particular the dilatation wave (velocity with zero-curl) and the distorted wave (velocity with zero-divergence) are discussed in Ref. 109. (See J. Larmor's footnote to *Stokes Papers* **5**, Appendix, Question 8 of 1854 Smith Prize, quoted in our Appendix.)

3. Maxwell's Work on Field Theory

Historically the three monumental papers (1855–1865) of Maxwell (1831–1879), which we shall refer to as 1, 2 and 3, founded the field theory of electromagnetism. For references, the three-volume set,[31–33] *The Scientific Letters and Papers of James Clerk Maxwell*, edited by P. M. Harman and the paper by C. N. Yang[141] are especially useful. Before we discuss Maxwell's papers in detail in this section, we shall first summarize what Faraday had done to inspire Maxwell.

3.1. *Michael Faraday (1791–1867)*

Faraday had become interested in electricity and magnetism after Örsted had discovered in 1820 that electric currents generate magnetism, and after Ampère had carefully and quantitatively studied the subject. Faraday's great contribution was the discovery of induction in 1831. Unable to describe the phenomena mathematically, he used his geometrical intuition to envisage space around a magnet filled by lines of force. There is a "tension" in the space which he called an "electrotonic state." Variation with time of the tension generates electricity, according to Faraday's ideas.

Among the many readable accounts of Faraday's life and work, besides Faraday's *Experimental Researches in Electricity* (Ref. 19) and *Faraday's Diary* (see Ref. 79), we mention (a) *Faraday as a Discoverer* by his friend J. Tyndall;[122] (b) *The Life & Letters of Faraday*, by H. B. Jones;[62] and (c) *Michael Faraday: A Biography*, by L. P. Williams.[134]

In Faraday's *Experimental Researches* (three volumes on electricity and one volume on chemistry and physics),[19] he recorded meticulously his work from 1831 to 1856. His private notes of the *Experimental Researches* contained entries of 16,041 paragraphs (see Ref. 122, p. 149).

3.2. *Faraday's work on electromagnetic induction*

On August 28, 1831, using two insulated copper coils wound on an iron ring, Faraday made the historic observation of the *transient* effect of the volta-electric induction. In a letter to his friend Richard Phillips (1778–1851, chemist and editor of *Philosophical Magazine*) dated September 23, 1831 (see Ref. 62), Faraday said:

> "*I am busy just now again on electromagnetism, and think I have got hold of a good thing, but can't say. It may be a weed instead of a fish that, after all my labour, I may at last pull up.*" (Faraday obviously remembered his previous unsuccessful attempts in 1824, 1825 and 1828.)

On October 17, 1831, Faraday made another historic discovery on the magneto-electric induction (see Fig. 2). In his diary entry on that date, Faraday wrote:

> "*A cylindrical bar magnet, $\frac{3}{4}''$ in diameter and $8\frac{1}{2}''$ in length, had one end just inserted into the end of the helix cylinder, then it was quickly thrust in the whole length, and* ***the galvanometer*** *needle moved; then pull out, and again the* ***needle moved but*** *in the opposite direction. This effect was repeated every time the magnet was put in or out, and therefore a wave of Electricity was so produced from* ***mere approximation of a magnet****, and not from its formation* ***in situ****.*" (*Emphasis*: original.)

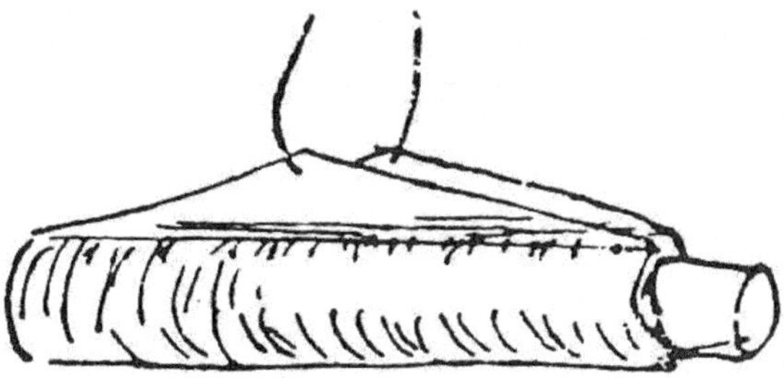

Fig. 2. A diagram from *Faraday's Diary* (October 17, 1831) (see Ref. 79). It shows a solenoid with coil attached to a galvanometer. Moving a bar magnet in and out of the solenoid generates electricity.

3.3. *Faraday's statements on the electrotonic state*

In his first paper on induction which appeared in the first series of Ref. 19, Vol. 1 and which was read before the Royal Society on November 24, 1831, Faraday pondered upon a new electrical state or condition of matter:

> "*Whilst the wire is subject to either volta-electric of magnetoelectric induction, it appears to be in a peculiar state; for it resists the formation of an electric current in it, whereas, if in its common condition, such a current would be produced; and when left uninfluenced it has the power of originating a current, a power which the wire does not possess under common circumstances. This electrical condition of matter has not hitherto been recognized, but it probably exerts a very important influence in many if not most of the phenomena produced by currents of electricity. For reasons which will be immediately apparent (paragraph* 71)*, I have, after advising with several learned friends, ventured to designate it as the* ***electrotonic*** *state.*" (paragraph 60)

> "*This peculiar state appears to be a state of tension, and may be considered as* ***equivalent*** *to a current of electricity, at least equal to that produced either when the condition is induced or destroyed.*" (paragraph 71)

In a letter to R. Phillips dated November 29, 1931 (cited in Ref. 62, Vol. 2, p. 9) in which Faraday summarized his latest discovery, he said:

> "*The effects are a current in the same direction when the induction is established; a reverse current when the induction ceases, and a* ***peculiar state*** *in the interim .... The condition of matter I have dignified by the term* ***Electrotonic, The Electrotonic State****. What do you think of that? Am I not a bold man, ignorant as I am, to coin words? But I have consulted the scholars.*"

In the same letter, Faraday offered an interesting observation:

"*It is quite comfortable to me to find that experiment need not quail before mathematics, but is quite competent to rival it in discovery.*"

Faraday was cognizant of the interplay between electricity and magnetism:

"*If we endeavour to consider electricity and magnetism as the results of two faces of a physical agent, or a peculiar condition of matter, exerted in determinable directions perpendicular to each other, then, it appears to me, that we must consider these two states or forces as convertible into each other in a greater or smaller degree; i.e. that an element of an electric current has not a determinable electric force and a determinable magnetic force constantly existing in the same ratio, but that the two forces are, to a certain degree, convertible by a process or change of condition at present unknown to us.*" (paragraph 1114)

Faraday asserted that the idea of electrotonic state should apply to the nonconductor as well:

"*Again and again the relation of conductors and nonconductors has been shown to be one not of opposite in kind, but only degree, and therefore, for this, as well as for other reasons, it is probable, that what will affect a conductor will affect an insulator also; producing perhaps what may deserve the term of the electrotonic state.*" (paragraph 1661)

In his 1852 paper *On Physical Lines of Magnetic Force*,[20,21] Faraday revisited his ideas of the electrotonic state and the physical lines of force:

"*I incline to the opinion that (the lines of magnetic force) have a physical existence correspondent to that of their analogue, the electric lines, and having that notion, am further carried on to consider whether they have a probable dynamic condition, analogous to the axis to which they consist in a state of tension round the electric axis, and may therefore be considered as static in their nature. Again and again the idea of an* **electrotonic** *state has been forced on my mind; such a state would coincide and become with that which would then constitute the physical lines of force.*" (paragraph 3269)

"*The motion of an external body, . . . , could not beget a physical relation such as that which the moving wire presents. There must, I think, be a previous state, a state of tension or a static state, as regards the wire, which, when the motion is superadded, produces the dynamic state or current of electricity. This state is sufficient to constitute and give a physical existence to the lines of magnetic force.*" (paragraph 3270)

It is interesting to point out that many years later, in 1881, in his Faraday Lecture before the Fellows of the Chemical Society of London, Helmholtz summarized Faraday's idea of the electrotonic state (see Ref. 41, pp. 401–436; Ref. 50, pp. 52–87).

> "(*Faraday*) *found that an electromotive force striving to produce these currents arises wherever and whenever magnetic force is generated or destroyed. He concluded that a part of space traversed by magnetic force there ought to exist a peculiar state of tension and that every change of this tension produces electromotive force.... This unknown hypothetical state he called provisionally the electrotonic state, and he was occupied for years and years in finding out what this electrotonic state was.*"

3.4. *Faraday's ideas on lines of force*

Faraday spoke of the metal cutting the magnetic curves (see first paper cited above, paragraph 114), and in a footnote, he visualized:

> "*By magnetic curves I mean the lines of magnetic force,... which would be depicted by iron filings, or those to which a very small magnetic needle would form a tangent.*"

Jones (Ref. 62, Vol. 2, p. 5) had this to say:

> "*This is the germ of those* ***lines of force****, which arose up in the mind of Faraday into* ***physical*** *and almost tangible matter. The influence which they had upon his thoughts and experiments will be seen from this time up to the date of the last researches which he sent to the Royal Society.*"

Twenty years after his 1831 discovery of electromagnetic induction, Faraday summarized his ideas in an important paper, "*On Lines of Magnetic Force.*"[20,21] In Faraday's own words, we read:

> "*It is also evident, by the results of the rotation of the wire and magnet* (paragraphs 3097 and 3106), *that when a wire is moving amongst equal lines* (*or in a field of equal magnetic force*), *and with an uniform motion, then the current of electricity produced is proportional to the* ***time****; and also to the* ***velocity*** *of motion.*" (paragraph 3114)
>
> "*They also prove, generally, that the quantity of electricity thrown into a current is directly as the amount of curves intersected.*" (paragraph 3115)

Faraday had the idea of superposition (paragraphs 3076 and 3082); he much preferred field lines to action at distance (paragraph 3074). The direction of the induced current relative to the field was discussed (paragraphs 114, 3079 and 3089).

Faraday clearly stated that the magnetic field lines are closed:

> "*Every line of force, therefore, at whatever distance it may be taken from the magnet, must be considered as a closed circuit, passing some part of its course through the magnet, and having an equal amount of force in the every part of its course.*" (paragraph 3117)

He went on to expand on the idea of flux:

> "*On the other hand, the use of the idea of* ***lines of force****, which I recommend, to represent the true and real magnetic forces, makes it very desirable that we should find a unit of such force, if it can be attainable.... In the mean time, for the enlargement of the utility of the idea in relation to the magnetic force, and to indicate its conditions graphically, lines may be employed as representing these units in any given case. I have so employed them in former series of these Researches* (paragraphs 2807, 2821, 2831, 2874, etc.), *where the direction of the* ***lines of force*** *is shown at once, and the relative amount of force, or of lines of force in a given space, indicated by their concentration or separation, i.e. by their number in that space.*" (paragraph 3122)

3.5. *Tribute to Faraday*

In the 1881 Faraday Lecture cited above at the end of Sec. 3.3, Helmholtz gave the following tribute:

> "*Now that the mathematical interpretation of Faraday's conceptions, regarding the nature of electric and magnetic forces has been given by J. C. Maxwell, we see how great a degree of exactness and precision was really hidden behind the words which to Faraday's contemporaries appear either vague or obscure; and it is in the highest degree astonishing to see what a large number of general theorems, the methodical deduction of which requires the highest powers of mathematical analysis, he found by a kind of intuition, with the security of instinct, without the help of a single mathematical formula.*"
>
> "*A single remarkable discovery may, of course, be the result of a happy accident and may not indicate the possession of any special gift on the part of the discoverer, but it is against all rules of probability that the train of thought which has led to such a series of surprising and unexpected discoveries as were those of Faraday should be without a firm, although perhaps hidden, foundation of truth.*"

Faraday was a kind and humble man, unconcerned with honors. He declined the offers of the presidency of the Royal Society in 1857, and the presidency of the Royal Institution in 1864, and a knighthood by Queen Victoria. Yet he treasured the love and sympathy of fellow man and prized it more than the renown which his science had brought him. After Faraday had read Tyndall's review of Faraday's *Experimental Researches* in the *Philosophical Magazine*, Tyndall recounted Faraday's remark to him:

> "*The sweetest reward of my work is the sympathy and good will which it has come to flow in upon me for all quarters of the world*" (Ref. 122, p. 150)

In a letter to Tyndall dated June 28, 1854, Faraday wrote (see Ref. 122, p. 152):

> "... *You are young, I am old.... But our subjects are so glorious that to work at this rejoices and encourages the feeblest, delights and enchants the strongest.*"

Maxwell's tribute to Faraday is cited toward the end of Sec. 9 in this paper.

3.6. *Maxwell's preparation*

Maxwell in his student days already was familiar with the concept of Potential Functions. In a "Manuscript on the Potential Function (Mathematical Theory of Polar Forces)" (circa 1851) (Ref. 31, pp. 210 and 211), Maxwell gave a then conventional view (apparently as a sophomore) of the potential:

> "*The most remarkable example of the use of a mathematical abstraction of this kind in mechanical science is the function used by Laplace* (*note added*: see Ref. 67, 1825) *in the theory of attractions and to which Green in his* (*note added*: see Ref. 23, 1828) *essay on Electricity has given the name of the Potential Function. We have no reason to believe that anything answering to this function has a physical existence in the various parts of space, but it contributes not a little to the clearness of our conceptions to direct our attention to the potential function as if it were a real property of the space in which it exists.*"

It is obvious from this paragraph that to introduce the potential function, which was considered *nonreal* at that time, required imagination and perhaps courage.

Upon his graduation from the Trinity College in 1854, Maxwell wrote to Thomson in February 20, 1854 (Ref. 31, No. 45, p. 237), asking for his advice toward readings in Electricity and specifically regarding how to follow in order the works of Ampère, Faraday and Thomson. Thomson apparently wrote a long letter to Maxwell on Electricity (not preserved), as acknowledged by Maxwell's letters to Thomson dated March 14 1854 and November 13, 1854. It is amazing how rapidly Maxwell has mastered the subject in eight months. In Maxwell's own words (Ref. 31, No. 51, p. 254):

> *Dear Thomson*:
>
> *I soon involved myself in that subject, thinking of every branch of it simultaneously, and have been rewarded of late by finding the whole mass*

of confusion beginning to clear up under the influence of a few simple ideas. ... I got up the fundamental principles of electricity of tension (note added by Harman: i.e. potential) easily enough. I was greatly aided by the analogy of the conduction of heat, which I believe is your invention at least I never found it elsewhere.... Now I have heard you speak of "magnetic lines of forces" (note added by Harman: Ref. 116) *and Faraday seems to make great use of them (note added by Harman:* Refs. 20 and 21), *but others seem to prefer the notion of attractions of elements of currents directly....*

Maxwell was aware of Weber work[124] through Thomson (see Ref. 31, No. 66, May 15, 1855).

It is interesting to note that, many years later, in his *Preface* to the first edition of *A Treatise on Electricity and Magnetism,*[88] Maxwell stated that

"... *before I began the study of electricity I resolved to read no mathematics on the subject till I had first read through Faraday's Experimental Researches on Electricity.*"

Clearly the ideas of Faraday's caught Maxwell's attention. One may surmise that in the period before his first great paper of 1855, Maxwell was trying hard to absorb both the physical intuitions of Faraday and the mathematical language of potential functions.

3.7. *Maxwell's Paper 1*[a]

Borrowing from Thomson's earlier papers on magnetism (see Subsec. 2.2 above) where the vector potential **A** was introduced (called three functions F, G and H), in paper 1, Maxwell *identified this vector potential with Faraday's intuitive idea of "electrotonic intensity."* In today's notation, Maxwell found

$$\mathbf{E} = -\dot{\mathbf{A}}\,, \tag{3.1}$$

which is Faraday's law. This is the most important result of Maxwell's paper 1.

With (3.1), Maxwell has expressed Faraday's law in the form of a partial differential equation, an achievement of the greatest importance. In his letter dated September 13, 1855 (Ref. 31, No. 71, p. 322), months before paper 1 was read at a meeting, Maxwell wrote:

Dear Thomson

"... *I intend next to apply to these facts Faraday's notion of an electrotonic state.* (*Harman's note*: *M. Faraday, Electricity,* **1**, 16 (Sec. 60). *I have worked a good deal of mathematical material out of this vein and I believe I have got hold of several truths which will find a mathematical expression in the electrotonic state....*"

[a]Paper 1 refers to the paper published in 1856, Ref. 80.

That Maxwell regarded his interpretation of Faraday's intuitive idea of an electrotonic intensity (which Maxwell sometimes called electrotonic function) as an important discovery is evident from the following passage at the end of paper 1, which indicated his misgivings that Thomson might take offense:

> "*With respect to the history of the present theory, I may state that the recognition of certain mathematical functions as expressing the "electrotonic state" of Faraday, and the use of them in determining electrodynamic potentials and electromotive forces is, as far as I am aware, original; but the distinct conception of the possibility of the mathematical expressions arose in my mind from the perusal of Prof. W. Thomson's papers....*"

We shall make the following remarks about paper 1:

(a) Paper 1 (which consists of two parts) is a long paper. Reprinted in Niven's volume,[95] it ran 75 pages from pp. 155–229. The important part of the volume is in the middle, from pp. 188–205, after the subtitle "Part II — On Faraday's 'Electrotonic State'."

(b) Maxwell adopted a left-handed x, y, z system (p. 194), and a left-handed definition of the operator curl (e.g. bottom of p. 194). He used Stokes' theorem repeatedly. (Please see the Appendix in this paper.)

(c) On p. 192, Maxwell explained that subscript 1 referred to magnetic quantities and subscript 2 to electric questions. Thus

$$\begin{aligned} a_1,\ b_1,\ c_1 &= \mathbf{B} \quad \text{magnetic induction}\,,\\ \alpha_1,\ \beta_1,\ \gamma_1 &= \mathbf{H} \quad \text{magnetic force}\,,\\ a_2,\ b_2,\ c_2 &= \mathbf{j} \quad \text{current}\,,\\ \alpha_2,\ \beta_2,\ \gamma_2 &= \mathbf{E} \quad \text{electromotive force}\,,\\ \alpha_0,\ \beta_0,\ \gamma_0 &= \mathbf{A} \quad \text{electrotonic intensity (or function)}\,. \end{aligned}$$

(d) On p. 195, Maxwell stated

> "*Our investigations are therefore for the present limited to closed currents; and we know little of the magnetic effects of any currents which are not closed.*"

In this connection, he wrote down explicitly the equation for closed currents, which is in today's notation:

$$\nabla \cdot \mathbf{j} = 0\,. \tag{3.2}$$

(e) In today's understanding, Faraday's law is expressed usually not as (3.1); but as

$$\oint \mathbf{E} \cdot d\boldsymbol{\ell} = -\iint \dot{\mathbf{H}} \cdot d\boldsymbol{\sigma}\,. \tag{3.3}$$

While the two are identical, (3.3) is certainly simpler. How did Maxwell arrive at the less simple form in this first attempt? To try to answer this question, it is important to realize that Faraday, in his intuitive discussions of the electrotonic state, was very far from such quantitative results as (3.1) or (3.3). It was Maxwell's great contribution to have arrived at (3.1). He did this in the four pages, 201–204 of paper 1.

First he integrated over the whole 3-space the equation

$$\nabla \cdot (\mathbf{A} \times \mathbf{H}) = \mathbf{H} \cdot (\nabla \times \mathbf{A}) - \mathbf{A} \cdot (\nabla \times \mathbf{H}) \tag{3.4}$$

obtaining

$$0 = \iiint \mathbf{H}^2 \, d\tau - \iiint \mathbf{A} \cdot \mathbf{j} \, d\tau \,. \tag{3.5}$$

He then wrote:

> "*Let us now consider the conditions of the conduction of the electric currents within the medium* ***during changes in the electrotonic state****. The method which we shall adopt is an application of that given by Helmholtz in his memoir on the Conservation of Force.* (*Note added*: Refs. 40 and 41)." (***Bold***: added.)

Finally using (3.5) he showed that Helmholtz's Conservation of Force (i.e. energy principle) led to

$$\iiint \mathbf{E} \cdot \mathbf{j} \, d\tau + \frac{d}{dt} \iiint \mathbf{A} \cdot \mathbf{j} \, d\tau = 0 \,. \tag{3.6}$$

He then derived (3.1) from (3.6).

(f) Maxwell had understood very well the gauge freedom in the definition of **A**, as is evident from his Theorem V (for which he gave credit to Thomson[114,115] for the modulo gradient term). So which of the gauge choices did he take in (3.1)? He did not enter into a discussion of this point in paper 1. We can assume, perhaps, that he realized that when integrated over a closed circuit the additional term in gauge freedom, being a gradient, would contribute nothing. So he did not comment on this freedom.

Maxwell remarked on p. 203 that it is better to use the vector potential which he called functions α_0, β_0, γ_0:

> "*We have now obtained in the functions* α_0, β_0, γ_0 *the means of avoiding the consideration of the quantity of magnetic induction which* ***passes through*** *the circuit. Instead of this artificial method we have the natural one of considering the current with reference to quantities existing in the same space with the current itself. To these I give the name of* ***Electrotonic functions, or components of the Electrotonic intensity****.*" (*Emphasis*: original.)

What did he mean by quantities "existing in the same space with the current itself"? Did he mean the vector potential can be chosen to be in the space where there is a current? Is this the reason that he did not comment on any gauge freedom?

3.8. *Maxwell's Paper 2*[b]

In paper 2 Maxwell introduced the displacement current, thereby reaching the historical conclusion that light is an electromagnetic wave. But what had led him to the displacement current is not clear. The paper had begun with the following declaration:

> "*My object in this paper is to clear the way for speculation in this direction, by investigating the mechanical results of certain states of tension and motion in a medium, and comparing these with the observed phenomena of magnetism and electricity.*"

There followed Part II — "Theory of Molecular Vortices applied to Electric Currents" with beautifully intricate lattices of vortex motion. To illustrate Maxwell's style of writing in his three monumental papers, we quote from paper 2:

> "*Let AB, Plate VIII,* p. 488, Fig. 2, *represent a current of electricity in the direction from A to B. Let the large, spaces above and below AB represent the vortices, and let the small circles separating the vortices represent the layers of particles placed between them, which in our hypothesis represent electricity.*"

which refers to the diagram reproduced here on the right side of our Fig. 3. How the diagram is related to electromagnetism was not explained by Maxwell.

Then, suddenly without warning, in Part III of the paper, Maxwell wrote:

> "*Prop XIV — To correct Eq.* (9) (*of Part I*) *of electric currents for the effect due to the elasticity of the medium.*"

The correction was to add the displacement current. With this historic addition Maxwell then proved charge conservation and wrote down as Eq. (113) what would be, in today's notation,

$$\nabla \cdot \mathbf{j} + \dot{\rho} = 0 \,. \tag{3.7}$$

He then derived the existence of electromagnetic waves.

Did Maxwell arrive at the concept of the displacement current through his complicated lattices of vortex motion? Through the elasticity of the medium? We cannot be sure. But it is worth noticing that Part III was published ten months after Part I, and eight or nine months after Part II, during which time he must have been working hard on the subject, arriving finally at the "correction." What had led to the "correction" we do not know, but he had written already at the beginning of Part I, months before he landed on the "correction,"

[b]Paper 2 refers to the papers published in 1861 and 1862, Refs. 81–85.

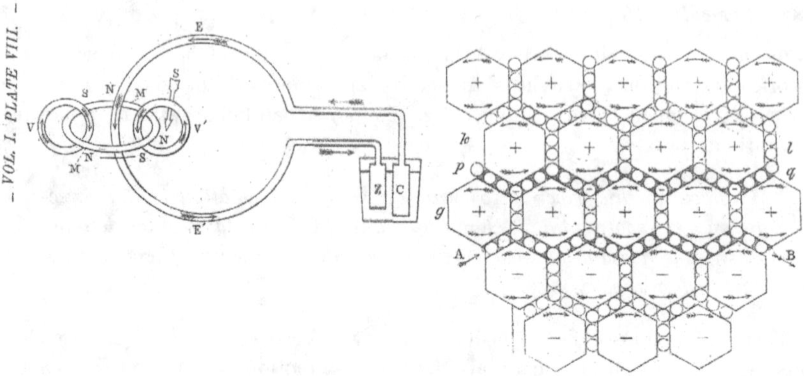

Fig. 3. A diagram in Maxwell's paper 2 as a model of the electromagnetic fields. This figure was also quoted in Ref. 6 with the following caption: Helmholtz (Ref. 44) describes it as "a systems of cells with elastic walls and spherical cavities... in which elastic balls can rotate and be flattened out by the centrifugal forces. In the wall of cells there must be other balls, of invariable volume, as friction rollers... their center of gravity... would merely be displaced by elastic yield of the center-wall... displacement of (the friction rollers) gives dielectric polarization of the medium; streaming of the same, as electric currents; rotation of the elastic balls corresponds to the magnetizing of the medium, the axis of rotation being the direction of the magnetic force."

> "...*I have found the geometrical significance of the 'Electrotonic State', and have shown how to deduce the mathematical relations between the electrotonic state, magnetism, electric currents, and the electromotive force, using the mechanical illustrations to assist the imagination, but not to account for the phenomena.*"

Thus he was constructing mechanical illustrations to "assist the imagination." Could it be that his lattices of vortex motion did assist him to incorporate the concepts of charge conservation and of the displacement current? If so, how?

But it is also possible that in the period after Part II of paper 2, Maxwell had decided to consider more than just "closed currents," and therefore had to use (3.7) rather than (3.2) (see (d) in Subsec. 3.7 above). It is easy to believe that once he embarked on this more general problem he would discover the displacement current.

We mention here in passing that for the mathematical backgrounds regarding the use of curl and divergence of vectors (in components), Maxwell cited the influence of Thomson[113–115] and Stokes.[108]

3.9. *Maxwell's Paper 3*[c]

Three years after the appearance of the last parts of paper 2 (Parts III and IV), Maxwell published paper 3 (which consists of six parts) with the title "A Dynamical

[c]Paper 3 refers to the paper published in 1865, Ref. 86.

Theory the Electromagnetic Field," in which he collected and tabulated the equations developed in the two earlier papers. It is especially noteworthy that he emphasized

> "*I have on a former occasion*[d] *attempted to describe a particular kind of motion and a particular kind of strain, so arranged as to account for the phenomena. In the present paper I avoid any hypothesis of this kind; and in using such words as electric momentum and electric elasticity in reference to the known phenomena, of the induction of currents and the polarization of dielectrics, I wish merely to direct the mind of the reader to mechanical phenomena which will assist him in understanding the electrical ones. All such phrases in the present paper are to be considered as illustrative, not as explanatory.*"

It is clear that Maxwell realized now that the mechanical models were irrelevant. It was the *field* that was the real essence. Also noteworthy was the following words in Section (74):

> "*In speaking of the Energy of the field, however, I wish to be understood literally. All energy is the same as mechanical energy, whether it exists in the form of motion or in that of elasticity, or in any other form. The energy in electromagnetic phenomena is mechanical energy. The only question is, Where does it reside? On the old theories it resides in the electrified bodies, conducting circuits, and magnets, in the form of an unknown quality called potential energy, or the power of producing certain effects at a distance. On our theory it resides in the electromagnetic field, in the space surrounding the electrified and magnetic bodies, as well as in those bodies themselves, and is in two different forms, which may be described without hypothesis as magnetic polarization and electric polarization, or, according to a very probable hypothesis as the motion and the strain of one and the same medium.*"

Here the fundamental spirit of the concept of the field, so dominant in the physics of the 20th century, was for the first time clearly and emphatically enunciated.

3.10. *Maxwell's statements about* **A** *being a momentum*

There were many places where Maxwell stated that **A** is a momentum. Some examples:

(a) In Part II of paper 2, we read (Ref. 95, Vol. 1, pp. 478 and 479)

> "*The electrotonic state, whose components are F, G, $H, \ldots$ correspond to the* ***impulse*** *which would act on the axle of a wheel in a machine*

[d]Here Maxwell referred to his paper 2.

if the actual velocity were suddenly given to the driving wheel, the machine being previously at rest....
This impulse may be calculated for any part of a system of mechanism, and may be called the ***reduced momentum*** *of the machine for that point.*" (*Emphasis*: original.)

(b) In Part III of paper 3 (Ref. 95, Vol. 1, p. 555) Maxwell stated that

Electromagnetic Momentum $\mathbf{A}$ = total impulse of the $\mathbf{E}$-field:

$$-\mathbf{A} = \int \mathbf{E}\, dt$$

or

$$\mathbf{E} = -\frac{d\mathbf{A}}{dt}.$$

(c) In Part III of paper 3, we read, "this total electromagnetic momentum is the same thing to which Professor Faraday has applied the name of the electrotonic state" (Ref. 95, Vol. 1, p. 556).

(d) In a more extended account given in his 1873 *Treatise* (Ref. 88, Vol. 2, articles 585–603), Maxwell stated quite clearly (article 590),

"*The vector* $\mathbf{A}$ *represents in direction and magnitude the time-integral of the electromotive force which a particle placed at the point* (x, y, z) *would experience if the primary current were suddenly stopped. We shall therefore call it the Electrokinetic Momentum at the point* (x, y, z). *It is identical with the quantity which we investigated in article* 405 *under the name of the vector-potential of magnetic induction.*"

4. The Action Principle

4.1. *Explicit versus implicit references to the expression* (p − eA)

Today, because of its relationship to covariant differentiation, the expression

$$\mathbf{p} - e\mathbf{A} \to -i\partial - e\mathbf{A}$$

has become a primary concept in physicists' description of the forces of nature. We want to trace in this section the early appearance in the literature of this fundamental expression.

4.2. *Explicit expression of* (p − eA)

One often quoted reference where the expression

$$\mathbf{p} = m\mathbf{v} + e\mathbf{A}$$

makes its appearance is the Max Born's 1923 Göttinger Lectures[7] discussing the Zeeman effect, as cited, e.g. in a Dirac's 1926 article.[15] J. H. Van Vleck (1926)[123] gave some interesting references to works on the modifications in the Bohr–Sommerfeld quantization to include relativity correction and magnetic fields. For example, the equivalent forms for the $(\mathbf{p} - e\mathbf{A})$ term appeared in the Hamilton–Jacobi formalism for the Zeeman effect treated by A. Sommerfeld[106] and by P. Debye[12] in 1916.

Among other references cited by Van Vleck, the earliest work where the expression $(\mathbf{p} - e\mathbf{A})$ can be explicitly found was George A. Schott's 1912 book, *Electromagnetic Radiation (and the Mechanical Reactions arising from it)*.[101] This book is the revised version of his 1908 Essay which was awarded the 1909 Adams Prize at the University of Cambridge. Additional materials beyond the 1908 Essay are contained in seven appendices. The expression $(\mathbf{p} - e\mathbf{A})$ is found in App. F, "*The Mechanics of the Lorentz Electron*." In the context of the Lorentz's relativistic mass correction (1895),[76] Schott used the Lagrangian written down in 1903 by K. Schwarzschild.[102–104] Schott wrote down the explicit Hamilton–Jacobi equations and the Hamiltonian, thereby exhibiting the explicit expression of $(\mathbf{p} - e\mathbf{A})$. So far, much to our surprise, we have not been able to find any explicit expression of $(\mathbf{p} - e\mathbf{A})$ prior to this work of 1912.

4.3. *Implicit expression of* (p − eA) *and the action principle*

On the other hand, one could ask instead the question: where was it *first* written down, the Lagrangian containing the interaction terms equivalent to $\mathbf{j} \cdot \mathbf{A}\, d^3x$, or $e\mathbf{v} \cdot \mathbf{A}$, which imply the expression $(\mathbf{p} - e\mathbf{A})$ in the Hamiltonian? To answer this question we have to examine the role of Thomson, Helmholtz and Larmor, etc. about their work on the Lagrangian and the action principle. We shall work our way backward in time.

4.3.1. *Joseph Larmor (1857–1942)*

W. Pauli[98] in his Handbuch article credited $(\mathbf{p} - e\mathbf{A})$ to Larmor's (1900b) book, *Aether and Matter*,[71] where for the Lagrangian for an electron moving with velocity $\mathbf{v}$, Larmor had included the magic term $\mathbf{v} \cdot \mathbf{A}$. Actually the relevant expressions in Larmor's 1900 book have already appeared in his earlier articles during 1895 to 1900[70,71] where the Lagrangian for a charged particle can be found. (See Eq. (4.1) below.)

Larmor wrote an article, "*On Least Action as the Fundamental Formulation in Dynamics and Physics*" in 1894. As a commentary on this article, Larmor gave an interesting "*Historical Note on Hamiltonian Action*" in a 1927 appendix to his collected papers (see Ref. 68, pp. 640–643):

> "*The original ideas of William Rowan Hamilton, nurtured, like all the Dublin school of his time, on the writers of the great age in France, have constituted an epoch in fundamental mathematical physics. They began with consideration of the properties of what he at first named on historical*

grounds the "Action function" for a System of Optical Rays, in a memoir communication in June 1824 *to the Royal Irish Academy when he was an undergraduate of nineteen years of age."* (*Hamilton's work appeared in Trans. R.I.A.* ***15*** (1828), *followed by three supplements in* 1830, 1831 *and* 1832 *respectively. A "magnificent synthesis," "On a General Method in Dynamics," appeared in two memoirs, Phil. Trans.* 1834, 1835).) *"After this date* (1835) *Hamilton seems to have parted company with fundamental dynamics and optics, devoting the remaining twenty-five years of his life largely to the domain of symbolic algebras, and particularly to the development, and the detailed application over wide ranges of subjects, of his system of quaternions, being the precursor in this extensive and imaginative field, on the future practical importance of which he laid great stress."*

Larmor then went on to trace the historical development of the action principle:

"The fundamental character of his (*Hamilton's*) *culminating doctrine of Varying Action, as implicating the whole range of physical science, seems to have been first emphasized for a wider audience in Thomson and Tait's classic book on Natural Philosophy* (1867). *It appears from the surviving letters that Tait had a large share of the merit of this recognition, by insistence on his more physical colleague's study of the work of his own Irish friend, of which the development, apart from applications in dynamical astronomy natural to the time, had hitherto, following the lead of Jacobi, been in the direction of abstract analysis of the implications of systems of partial differential equations.... From Thomson and Tait the illumination passed on to Helmholtz, culminating in his* [1886] *extensive memoir*[45] *Über die physikalische Bedeutung des Princips der kleinsten Wirkung...."*

Here Larmor regarded his own 1894 paper on Least Action as of "preliminary and analogical scope as contrasted with Helmholtz's bolder formulations developed soon afterwards."

Larmor in fact gave an interesting prehistory of the action principle:

"As a matter of scientific history, a germ of the earlier Hamiltonian developments for rays may be found, already in Newton's days, in the principle of Cotes, expanded by R. Smith in his Systems of Opticks and recalled to modern attention by Rayleigh. It asserts that the apparent distance of an object at A, seen across any optical instrument by an observer at any other place B, is equal to the apparent distance of the same object located at B as seen from A. For the special case when A and B are conjugate foci, the relations of object and image, as earlier generalized for all systems by Huygens, emerge from this principle...."

In a footnote added in 1927 to his 1895 paper in his collected papers (Ref. 69, p. 414), Larmor acknowledged the influence of Thomson's work on dynamics:

> "*The main idea of these three memoirs was to express and extend the electrical theory by reference, under direct inspiration mainly from Lord Kelvin's writing, to a guiding working model framed as far as possible directly on dynamical principles....*"

Larmor's 1895 (see Ref. 70, p. 568) version of the kinetic energy of a moving electron (in slightly modified notation) reads:

$$T = \frac{1}{2}m\mathbf{v}^2 + e\mathbf{v}\cdot\mathbf{A} + \cdots . \tag{4.1}$$

4.3.2. *William Thomson (1824–1907)*

Thomson and Tait's *Treatise on Natural Philosophy*[117] gave "an account of the splendid dynamical theory founded by d'Alembert and Lagrange." In article 310, one finds the equations of impulsive motion:

> "*The effect of any stated impulses, applied to a rigid body, or to a system of material points or rigid bodies connected in any way, is to be found most readily by the aid of d'Alembert's principle (note added: in* 1742) *according to which the given impulses, and the impulsive reaction against the generation of motion, measured in amount by the momenta generated, are in equilibrium; and are therefore to be dealt with mathematically by applying to them the equations of equilibrium of the system.*"

Thomson and Tait of course did not have the expression $(\mathbf{p}-e\mathbf{A})$. But their work was important historically because Maxwell (i) attributed his dynamics theory to Thomson–Tait's treatment of the impulsive forces (Ref. 88–90, Vol. 2, article 554); and (ii) appreciated from his study of Faraday's law of induction that the electrotonic state (namely, the vector potential) is an impulse. (See Subsec. 3.10 above.)

There are several dynamical examples discussed in Thomson–Tait's *Natural Philosophy* which contain linear terms in the velocities (besides the quadratic terms). This would give rise to a difference between the *canonical* momentum $\mathbf{p}$ and the *kinematical* momentum $m\mathbf{v}$, and such difference could be interpreted as that due to the existence of a vector potential.

4.3.3. *Hermann von Helmholtz (1821–1894)*

During 1886–1892, Helmholtz wrote three papers on the principle of least action. The 1892 paper has the Lagrangian with the $\mathbf{v}\cdot\mathbf{A}$ term, and the difference between the canonical momentum and the kinematical momentum is clearly recognized. Thus Helmholtz's work preceded Larmor's by some two to eight years. Indeed, in his 1894 paper, Larmor acknowledged (see Ref. 73, p. 447):

> "*It may be mentioned that a scheme for expressing the equations of electrodynamics by a minimum theorem analogues to the principle of Least Action has recently been constructed by von Helmoltz* (1892).[47,48]"

A. Einstein (1879–1955) was deeply interested in electrodynamics since the age of 16.[100] As revealed by his letters to M. Maric,[100] Einstein was expressing both his admiration for and frustration with Helmholtz's work:

(a) Early August 1899 letter:

> "*I admire the originality and independence of Helmholtz's thought more and more.*"

(b) 10(?) August 1899 letter:

> "*I returned the Helmholtz*[50] *volume and am now rereading Hertz's*[54,55] *propagation of electric forces with great care because I didn't understand Helmholtz's*[47] *treatise on the principle of least action in electrodynamics. I'm convinced more and more that the electrodynamics of moving bodies as it is presented today doesn't correspond to reality, and that it will be possible to present it in a simple way.*"

On the other hand, perhaps not surprisingly, Oliver Heaviside (Ref. 38, Vol. 2, p. 509) had a very disparaging opinion on the least action principle:

> "*Finally, there is the Principle of Least Action. Now, Least Action has no more to do with the matter than the man in the moon, as far as I can see. It is quite unnecessary, to begin with. Next, it obscures and complicates the matter, so much so as sometimes to lead to serious error. I made this remark advisedly, remembering previous applications of the Principle of Least Action to electromagnetics, which is much clearer without it.*"

4.3.4. *William Rowan Hamilton* (*1805–1865*)

Hamilton's work was briefly alluded to in Subsec. 4.3.1 above. It is interesting to trace how Hamilton's idea on the least action principle evolved from geometrical optics to dynamics. Hamilton[24] dealt with the properties of systems of rays under reflection. Hamilton[25] treated refraction. Here he derived the Euler–Lagrange equations for the rays from the principle of least action (generalizing Fermat's principle of least time). We note that Fermat (1601–1665) stated his least-time principle first for mirror reflection. Huygens (1629–1695) verified Fermat principle for refraction. Pierre-Louis Moreau de Maupertuis (1698–1759) stated the least action principle in 1744. Euler (1707–1793) had his least action (in the form of integral of $v\,ds$) principle using calculus of variation in 1744. Subsequently Lagrange (1736–1813) applied Euler's principle of least action to problem in dynamics.

Hamilton[26] contained the interesting observation:

> "*Meanwhile, it appears that if a general method in deductive optics can be attained at all, it must flow from some law or principle, itself of the highest generality, and among the highest results of induction.... The answer, I think, must be, the principle or law, called usually the Law of*

Least Action;... that this linear path of light, from one point to another, is always found to be such, that if it be compared with the other infinitely various lines by which in thought and in geometry the same two points might be connected, a certain integral or sum, called often Action, and depending by fixed rules on the length, and shape, and position of the path, and on the media which are traversed by it, is less than all the similar integrals for the other neighboring lines, or, at least, possesses, with respect to them, a certain stationary property.... From this Law, then, which may perhaps, be named the Law of Stationary action, it seems that we met most fitly and with best hope set out, in the synthetic or deductive process, and in search of a mathematical method."

In paragraphs following the above quote, Hamilton noted that

"Newton, however, by his theory of emission and attraction, was led to conclude that the velocity of light was directly, not inversely, as the index, and that it was increased instead of being diminished on entering a denser medium; a result incompatible with the theorem of shortest time in refraction. The theorem of shortest time was accordingly abandoned by many, and among the rest by Maupertuis, who, however, proposed in its stead, as a new cosmological principle, the celebrated law of least action which has since acquired so high a rank in mathematical physics, by improvements of Euler and Lagrange."

Hamilton[27] was the turning point on the application to dynamics of a general mathematical method previously applied to optics, while Hamilton[28] and Hamilton[29] gave the fully developed general method in dynamics.

4.4. *Little emphasis on the Hamiltonian and* $(\mathbf{p} - e\mathbf{A})$ *until quantum theory*

The history of post-Maxwell electromagnetism is described in many books, such as *The Maxwellians* by Hunt;[60] *From Maxwell to Microphysics* by Buchwald;[8] and *Intellectual of Mastery of Nature*: *Theoretical Physics from Ohm to Einstein* by Jungnickel and McCormmach.[63,64] The two encyclopedia articles by Lorentz[77,78] give a technical summary of the development up to and around the turn of the century. O. Darrigol[11] has given a historical account, *Electrodynamics from Ampère to Einstein*.

From our discussion of the history of $(\mathbf{p}-e\mathbf{A})$, we see that while the least action principle was invoked by a number of people, it appears that not much emphasis was placed on the use of the Hamiltonian $H = (\mathbf{p} - e\mathbf{A})^2/2m$ before the advent of quantum theory and quantum mechanics. This fact had been explicitly stated in Whittaker[133] in his article on Hamilton.

5. Heaviside–Hertz's Elimination of A

5.1. *Heaviside's view on the role of the potentials*

O. Heaviside (1850–1925) seemed to have derived great satisfaction in 1885 in eliminating the vector potential **A** from the Maxwell equations. Disagreeing with Maxwell's elevation of **A** to the rank of a fundamental quantity, Heaviside regarded "**A** and its scalar potential *parasite* ψ sometimes causing great mathematical complexity and indistinctiveness; and it is, for practical reasons, best to *murder* the whole lot, or, at any rate, merely employ them as subsidiary functions." (Ref. 35 or 36, Vol. 2, p. 482.) "Thus ψ and **A** are murdered, so to speak, with a great gain in definiteness and conciseness." (Ref. 35 or 36, Vol. 2, p. 468.)

In a PostScript to the 1889 *Phil. Mag.* article referred to above, Heaviside stated,

> "*According to the way of regarding the electromagnetic quantities I have consistently carried out since January 1885, the question of the propagation of, not merely the electrical potential* ψ*, but the vector potential* **A***, does not present itself as one for discussion; and, when brought forward, proves to be one of a metaphysical nature.*"

In his Preface to his two-volume set of *Electrical Papers*,[36] Heaviside reiterated,

> "*The duplex method (referring to his recasting of the Faraday's law as* curl **E** *in parallel to the* curl **B** *for the Ampère's law as Maxwell has it) eminently suited for displaying Maxwell's theory, and brings to light many useful relations which were formerly hidden from view by the intervention of the vector potential and its parasites.*"

In his Preface to his three-volume set of *Electromagnetic Theory*,[37–39] again he stated, referring to "an outline scheme of the fundamentals of electromagnetic theory from the Faraday–Maxwell point of view, with some small modifications and extensions upon Maxwell's equations,"

> "*... it is also done in the duplex form I introduced in* 1885*, whereby the electric and magnetic sides of electromagnetism are symmetrically exhibited and connected, whilst the 'forces' and 'fluxes' are the objects of immediate attention, instead of the potential functions which are such powerful aids to obscuring and complicating the subject, and hiding from view useful and sometimes important relations.*"

In a similar vein, Heaviside (Ref. 37, Vol. 1, p. 383) has this to say:

> "*The reader who is acquainted with the (at present) more 'classical' method of treating the electromagnetic field in terms of the vector and scalar potentials cannot fail to be impressed by the difference of procedure and of ideas involved. In the present method we are, from first to last, in contact with those quantities which are believed to have physical significance (instead of*

with mathematical functions of an essentially indeterminate nature), and with the laws connecting them in simplest form."

What did Heaviside mean by "indeterminate nature"? Whatever he meant, we observe that *the flexibility of the potential functions is precisely the concept essential to today's understanding of the fundamental forces of nature. This flexibility is what is meant by the technical term: gauge invariance.*

5.2. *Hertz's view on the role of the potentials*

H. Hertz (1857–1894) wrote, among others, three interesting theoretical papers on electrodynamics. According to Hertz (Ref. 51, footnote, p. 289), Helmholtz in 1847 and Thomson in 1848 deduced induction from the electrodynamic action.

Hertz (Refs. 51 and 59, p. 286) did away with the vector potential:

"*The vector potentials of electric and magnetic currents have hitherto occurred as quite separate, and from them the electric and magnetic forces were deduced in an unsymmetric manner. This contrast between the two kinds of forces disappear as soon as we attempt to determine the propagation of these forces themselves, i.e. as soon as we eliminate the vector potentials from the equation...*"

Hertz (Refs. 53 and 55, p. 209) regarded the potentials as *scaffolds*:

"*Als eine rudimentare Erscheinung mathematischer Natur nenne ich das Vorherrschem des Vectorpotentiales in den Grundgleichchungen. Bei dem Aufbau der neuren Theorien dienten die Potentiale als Gerüst... Nachdem wir aber gelernt haben, die Kräfte welber als Grössen der letzteren Art anszusehen, hat ihr Ersatz durch Potentiale nur dann einen Zweck, wenn damit ein mathematischer Vorteil erreicht wird. Und ein solcher scheint mir mit der Einführung des Vectorpotentiales in die Grundgleichungen nicht verbunden, in whelchen man ohnehin erwarten darf, Beziehungen zwischen Grössen der physikalischen Beobachtung, nicht zwischen Rechnungsgrössen zu finden.*"

Hertz (Refs. 53 and 55, pp. 209 and 210) acknowledged that Heaviside had worked along the same direction (in eliminating the vector potential) since 1885.

5.3. *Maxwell*

Less well known than the elimination of the vector potential in the hands of Heaviside and Hertz, Whittaker (Ref. 132, p. 287) asserts that Maxwell himself did it in 1868:

"*As the potentials do not possess any physical significance, it is desirable to remove them from the equations. This was done afterwards by Maxwell himself, who in* 1868 *proposed to base the electromagnetic theory of light solely on the equations*

$$\text{curl}\,\mathbf{H} = 4\pi\mathbf{S}\,, \quad \text{curl}\,\mathbf{E} = -\frac{\partial\mathbf{B}}{\partial t}$$

together with the equations which define **S** *in terms of* **E**, *and* **B** *in terms of* **H**." (See Ref. 87.)

6. A Cannot be Eliminated: Aharonov–Bohm Experiment

Although Maxwell himself in all of his papers and in his *Treatise* always explicitly used the vector potential in his equations, throughout most of 20th century the Heaviside–Hertz form of Maxwell's equations were taught to college students all over the world. The reason is quite obvious: the Heaviside–Hertz form is simpler, and exhibits an appealing near symmetry between **E** and **H**.

With the widespread use of this vector-potential-less version of Maxwell's equations, there arouse what amounted to a dogma: that the electromagnetic field resides in **E** and **H**. Where both of them vanish, there cannot be any electromagnetic effects on a charged particle.

This dogma explains why when the Aharonov–Bohm[1] article was published it met with general disbelief. A year later, Chambers[9] claimed to have verified experimentally the *A–B* effect. But he used a tapered magnetic needle instead of a long solenoid, and the leakage magnetic field led to many confusing discussions about whether he had really seen the effect. Fortunately, beautiful quantitative experiments were finally performed by Tonomura and collaborators,[120,121] showing conclusively the existence of the *A–B* effect, which means that **E** and **H** together do not completely describe the electromagnetic field, and that the vector potential cannot be totally eliminated in quantum mechanics. In the words of Wu and Yang:[136] the field strengths *underdescribe* electromagnetism.

7. Gauge Theory

The history of gauge theory has been described in detail in several publications in the last 30 odd years.[13,141,143–145,147–150,96,97,61] We shall only outline here the main conceptual steps in this history.

7.1. *H. Weyl (1918)*

In 1918 and 1919 three papers appeared that served to start in time a great transformation of fundamental physics in the latter part of the 20th century. Yet these papers were not written by a physicist, nor were they based on experimental facts. And they were immediately objected to by Einstein. The author was H. Weyl (1885–1955), a distinguished mathematician who liked to dabble in philosophy. What were his motivations? How did he land on the chief ideas behind these papers?

We do not know the full answers to these questions. But we shall make a few remarks as follows.

(1) Weyl was very much motivated by Einstein's geometrization of gravitational forces through a quadratic infinitesimal form ds^2. He wanted to geometrize electromagnetic interactions through a linear form $d\varphi = \varphi_\mu \, dx^\mu$.

(2) He was impressed by the ideas of Levi-Civita, Hessenberg and himself about the parallel displacement of a vector in Riemannian geometry. He asked, if after going around a closed loop in space–time, the parallel displaced vector may not return to its original direction, why not also its length? ("Warum nicht auch seine Länge?") (See Ref. 145, footnote 14.)

He seemed to try to apply this length-changing idea by making a conformal dilation which varies from point to point in space–time. To this end he introduced in the middle of his first paper[125] a "Proportionalitätsfaktor"

$$\exp\left(\int d\varphi\right), \quad d\varphi = \varphi_\mu \, dx^\mu = \boldsymbol{\phi} \, d\mathbf{x}$$

in which he later identified φ_μ with the vector potential A_μ times a numerical constant. Thus we read in that paper

$$(ds')^2 = \lambda (ds)^2 \,, \quad d\varphi' = d\varphi - \left(\frac{d\lambda}{\lambda}\right), \tag{7.1}$$

where λ is the conformal factor. These two equations later evolved, in 1929, into the gauge transformations of the first and second kind, by which time he had also adopted the insertion of the factor $i = \sqrt{-1}$ in the numerical constant relating φ_μ with A_μ. He wrote in 1929:[74]

> "*By this new situation, which introduces an atomic radius into the field equations themselves — but not until this step — my principle of gauge-invariance, with which I had hoped to relate gravitation and electricity, is robbed of its support. But it is now very agreeable to see that this principle has an equivalent in the quantum-theoretical field equations which is exactly like it in formal aspect; the laws are invariant under the simultaneous replacement of* φ *by* $e^{i\lambda}\varphi$, ϕ_α *by* $\phi_\alpha - \partial\lambda/\partial x_\alpha$, *where* λ *is an arbitrary real function of position and time. Also the relation of this property of invariance to the law of conservation of electricity remains exactly as before… the law of conservation of electricity*
>
> $$\frac{\partial s_\alpha}{\partial x_\alpha} = 0$$
>
> *follows from the material as well as from the electromagnetic equations. The principle of gauge-invariance has the character of general relativity since it contains an arbitrary function* λ, *and can certainly only be understood in terms of it.*"

Please notice that he has changed notation here. The new λ is the old $(\ln\lambda)$, and the old conformal factor is now a phase factor, multiplied not onto the metric ds^2, but onto the wave function ψ.

Thus the key ideas of Weyl were already in his first paper.[125] The latter two papers of 1918–1919, for our present investigation of the evolution of ideas, did not really add anything of importance.

Weyl's papers on physics were rambling, discursive and philosophical. They were also extraordinarily original. He was exploring new ideas, so there was great fluidity in his style. He frequently changed names for key concepts. Proportionalität later became Strecke, so that when Schrödinger in 1922 referred to the Proportionalitätsfaktor, it became Streckenfaktor (see Yang Ref. 146).

When the concept of gauge invariance first came up, he called it Massstab-Invarianz. But later he settled on Eichinvarianz. English translation of this term is now gauge invariance, but had been at times calibration invariance and measure invariance.

(3) Why did Weyl first consider a Proportionalitätsfaktor, and not a phase factor? The answer is simple: (a) Weyl had started his thinking with Riemannian geometry, not waves. So there were only real numbers. (b) The importance of phase factors in physics only began with de Broglie and Schrödinger a few years later when wave mechanics was born.

(4) Weyl strongly associated gauge-invariance (measure-invariance) with the conservation of electricity. A paragraph in Ref. 126 (in the 1923 translation) reads:

> "*For we shall show that as, according to investigations by Hilbert, Lorentz, Einstein, Klein, and the author, the four laws of the conservation of matter (the energy–momentum tensor) are connected with the invariance of the action quantity (containing four arbitrary functions) with respect to transformations of coordinates, so in the same way the law of the conservation of electricity is connected with the 'measure-invariance'.*"

(5) Starting from the Proportionalitätsfaktor there arose an operator

$$(\boldsymbol{\partial} - e\mathbf{A})$$

which played an important part in Weyl's theory. In today's terminology, Weyl was dealing with connections on a R_1 bundle where the fiber is the noncompact additive group of real numbers.

7.2. *Einstein's objection*

Einstein objected to Weyl's idea in a postscript published in 1918 together with Ref. 126. He said if Weyl's idea were correct,

> "*The length of a common ruler (or the speed of a common clock) would depend on its history.*" (Our translation.)

which was deeply perspective and killed the original version of Weyl's idea.

As we shall see below, the idea was later resurrected in 1927–1929 by the insertion of a factor $i = \sqrt{-1}$ in Weyl's Proportionalitätsfaktor, converting it to a phase factor: a change of the quantum mechanical phase factor does not affect the length of a ruler, so Einstein's objection was rendered nonoperative.

It was pointed out in a paper by Yang in 1983 (Ref. 144) that although Einstein's objection is nonoperative in the sense that the phase factor does not affect the length of a ruler, but one could raise the question whether the phase, or difference between phases, can be observed. Had Einstein or Weyl revisited in 1929 the original Einstein's objection, the *A–B* effect would have been proposed some thirty years before Aharonov and Bohm came upon it in 1959.

7.3. *V. Fock (1898–1974) and F. London (1900–1954) (1926 and 1927)*

After quantum mechanics was developed in 1925 and 1926, Fock and London independently pointed out that in quantum mechanics,

$$(\mathbf{p} - e\mathbf{A}) \to (-i\boldsymbol{\partial} - e\mathbf{A}) = -i(\boldsymbol{\partial} - ie\mathbf{A})$$

which is very similar to Weyl's $(\boldsymbol{\partial} - e\mathbf{A})$ except for the replacement of $\mathbf{A}$ with $i\mathbf{A}$. Of course with this replacement, Weyl's factor becomes

$$\exp \int [-ie\mathbf{A}\, d\mathbf{x}]$$

which is a phase factor. Thus "gauge theory" is a misnomer. It should have been called a "phase theory."

7.4. *H. Weyl (1929)*

Weyl came back in 1929 in a paper that incorporated the factor $i = \sqrt{-1}$. It is a very remarkable paper which for the first time explicitly discussed the concept of gauge transformations.

We had mentioned above in Subsec. 7.1 that Weyl had strongly associated gauge invariance with charge conservation. This association found its way into Pauli's review article of 1941 which was the "bible" in field theory for the graduate students of the 1940's. We shall see in Sec. 8 that this association turned out to be important for later developments.

Throughout the 1930's and 1940's gauge invariance was well known to theoretical physicists, but it produced nothing really new. Its only use seemed to be for a young theorist to spring a "smart" question at the end of a seminar: "Is your result gauge invariant?"

8. A Principle for Interactions: Yang and Mills (1954); Symmetry Dictates Interaction

The discovery of many "strange" particles in the late 1940's and early 1950's was the background motivation for finding a general *principle* for interactions between particles. (See Ref. 143, p. 19.) In 1954 Yang and Mills published a short abstract[138] and a short paper[139] generalizing Weyl's gauge invariance to non-Abelian gauges. It is interesting today to see how the authors stated the motivation for this generalization.

We reprint here the full text of the short abstract:

"***Isotopic Spin Conservation and a Generalized Gauge Invariance***

The conservation of isotopic spin points to the existence of a fundamental invariance law similar to the conservation of electric charge. In the latter case, the electric charge serves as a source of electromagnetic field; an important concept in this case is gauge invariance which is closely connected with (1) *the equation of motion of the electromagnetic field,* (2) *the existence of a current density, and* (3) *the possible interactions between a charged field and the electromagnetic field. We have tried to generalize this concept of gauge invariance to apply to isotopic spin conservation. It turns out that a very natural generalization is possible. The field that plays the role of the electromagnetic field is here a vector field that satisfies a non-linear equation even in the absence of other fields.* (*This is because unlike the electromagnetic field this field has an isotopic spin and consequently acts as a source of itself.*) *The existence of a current density is automatic, and the interaction of this field with any fields of arbitrary isotopic spin is of definite form* (*except for possible terms similar to the anomalous magnetic moment interaction terms in electrodynamics*)."

The focus here is to look for a *principle* for interactions through using conserved quantities as sources of vector transmitting fields.

In the paper, however, which was submitted a few months later, a different motivation was summarized in the first sentence of the abstract of the paper:

"*It is pointed out that the usual principle of invariance under isotopic spin rotation is not consistent with the concept of localized fields.*"

This sentence is certainly overstated: it would be better to read

"*It is pointed out that the usual principle of invariance under isotopic spin rotation is not consistent with* ***the spirit of*** *the concept of localized fields.*"

Be that as it may, for the purpose of the present paper, the important point about the generalization was the replacement of $(\partial - ie\mathbf{A})$ with $(\partial - ie\mathbf{B})$ where $\exp(-ie\mathbf{B})$ is a matrix representation of a Lie group.

Gauge Field Terminology	Bundle Terminology
Gauge (or global gauge)	principal coordinate bundle
Gauge type	principal fiber bundle
Gauge potential b_μ^k	connection on a principal fiber bundle
S	transition function
Phase factor Φ_{QP}	parallel displacement
Field strength $f_{\mu\nu}^k$	curvature
Source (electric) J_μ^k	?
Electromagnetism	connection on a U_1 bundle
Isotopic spin gauge field	connection on a SU_2 bundle
Dirac's monopole quantization	classification of U_1 bundle according to first Chern class
Electromagnetism without monopole	connection on a trivial U_1 bundle
Electromagnetism with monopole	connection on a nontrivial U_1 bundle

Fig. 4. "Dictionary" in the Wu–Yang paper.[136] What was particularly interesting to the mathematicians was the ? mark, i.e. the concept of the electric source 4-vector J_μ, which was among the basic concepts in physics. This is a good example of how a fundamental and subtle mathematical concept could be discovered by very different routes, in mathematics and in physics, through very different traditions and very different value judgments. Please compare with Ref. 149.

Thus given a Lie group associated with a conservation law, i.e. an invariance, the new principle of interaction gives rise to a gauge theory that incorporates that invariance, in a beautiful and unique way. Invariance, of course, means symmetry, hence the statement *symmetry dictates interaction* (C. N. Yang, *Phys. Today* **33**, 42 (1980)).[142]

This gauge theory of interactions, however, did not find acceptance among the physicists in the 1950's and in much of the 1960's, despite its appealing elegance. The reason is the bothering question about whether the vector field in the theory should have mass zero. It was only in the late 1960's that the idea of *symmetry breaking* was applied to non-Abelian gauge theories to remove this bothersome question. This development led to the *Standard Model*, which has been spectacularly successful in describing essentially all aspects of particle physics in the last 30 years. The exciting history of the *standard model*, however, lies outside of the considerations of the present paper.

9. Connections on Fiber Bundles

The mathematical meaning of the matrix $-ie\mathbf{B}$ in $(\boldsymbol{\partial} - ie\mathbf{B})$ as a "connection" was not understood by Yang and Mills in the 1950's. In 1983 Yang described (Ref. 143, p. 73) how during 1967–1968 he finally noticed the similarity between the nonlinear terms in the Yang–Mills paper and the nonlinear terms in the definition of the Riemann curvature tensor, and thereby began to understand the geometrical meaning of gauge theory. He then learned the rudiments of fiber bundle theory from his mathematician colleague Jim Simons, and wrote a paper in 1975 with T. T. Wu[136] which contained a "dictionary" of the corresponding concepts in physics and in mathematics (Fig. 4). They discussed this "dictionary" with I. M. Singer when the latter visited Stony Brook in the summer of 1976. Singer then lectured on gauge theory at Oxford and wrote an article with Atiyah and Hitchin. He later described this experience in 1988:[105]

> "*Thirty years later, I found myself lecturing on gauge theories at Oxford, beginning with the Wu and Yang dictionary and ending with instantons, i.e. self-dual connections. I would be inaccurate to say that after studying mathematics for thirty years, I felt prepared to return to physics. Instead, elementary particle physics turned to modern mathematics, and some of us found the interplay full of promise.*"

With the 1978 article by Atiyah, Hitchin and Singer,[5] an intense period of interest among mathematicians in gauge theory began.

10. The Concepts of Connections and Fiber Bundles in Mathematics

The concept of connections in mathematics had started with the works of Riemann, Christoffel, Ricci, Levi-Civita and E. Cartan. (We have not studied this evolution in detail.) For obvious reasons the concept had started with the geometry of the tangent bundle of a manifold. Extension to other bundles took place in the 20th century. According to Dieudonné (Chapter III of Ref. 14), this was first done in a paper by Hotelling in 1925. The word fiber (Faser) later appeared in the work of Seifert in 1932. Then, "*The first genuine 'fiber space' was only defined by Hassler Whitney in* 1935 *under the name 'sphere-space'.*"

According to Spivak,[107] connections on principle bundles were first developed in Ehresmann.[17]

The field of global differential geometry, which had profoundly influenced every major branch of mathematics in the latter half of the 20th century, was ushered in by the works of Chern. (See Ref. 14.)

It is interesting to note here that an example of a nontrivial bundle had already occurred in Dirac's 1931 paper[16] on magnetic monopoles. (See also the paper by Wu and Yang, Ref. 137.)

Also interesting is the fact that Maxwell was keenly aware of the possible origin of deep mathematics in the examination of the physical world. Some 50 years before Maxwell was born, Lagrange (1736–1813) had written in 1781 to his friend and mentor d'Alembert (1717–1783), lamenting the dearth of new developments in mathematics and in mechanics:

> *"I also think that the mine has become too deep and sooner or later it will be necessary to abandon it if new ore-bearing veins shall not be discovered. Physics and chemistry display now treasures much more brilliant and easily exploitable, thus, apparently, everybody has turned completely in this direction, and possibly posts in geometry in the Academy of Sciences will some day be like chairs in Arabic Language in universities at present."*

(See Arnold in Ref. 4, and the discussion by Yang in Ref. 149).

In the event, "new ore-bearing veins" did get discovered — by Faraday (1791–1867), whose ideas led Maxwell to his equations and to the evolution of concepts outlined in the present paper. Faraday knew little mathematics. He was a great experimentalist, but had profound geometrical intuition. He conceived of the "lines of force" and "the electrotonic state." He had results expressed neither in the form of the differential equation

$$\nabla \times \mathbf{E} = -\dot{\mathbf{H}} \tag{10.1}$$

nor in the form of the integral equation

$$\oint \mathbf{E} \cdot d\ell = -\iint \dot{\mathbf{H}} \cdot d\sigma \,. \tag{10.2}$$

Rather his results were in terms of concepts *equivalent* to this integral equation, and Maxwell was the one who understood this equivalence. In the preface to the first edition of *Treatise*, published years later, Maxwell characterized Faraday's intuitive results as follows:

> *"As I proceeded with the study of Faraday, I perceived that his method of conceiving the phenomena was also a mathematical one, though not exhibited in the conventional form of mathematical symbols....*
>
> *When I had translated what I considered to be Faraday's ideas into a mathematical form, I found that in general the results of the two methods coincided, so that the same phenomena were accounted for, and the same laws of action deduced by both methods, but that Faraday's methods resembled those in which we begin with the whole and arrive at the parts by analysis, while the ordinary mathematical methods were founded on the principle of beginning with the parts and building up the whole by synthesis...."*

Maxwell was really saying here that Faraday had used the integral form (10.2), and ordinary mathematical methods had used the differential form (10.1), of the same physical law.

The equivalence of the two was via Stokes' theorem (see Appendix) which today has evolved into a key idea in differential geometry, assuming the wonderfully simple and potent form:

$$\int_{\partial R} \mathbf{A} = \int_R d\mathbf{A} \,.$$

Upon Faraday's death on August 25, 1867, Maxwell wrote a tribute which was published in *Nature*. Two passages in this tribute were as follows:

> "*The way in which Faraday made use of his idea of lines of force in co-ordinating the phenomena of magnetoelectric induction shews him to have been in reality a mathematician of a very high order — one from whom the mathematicians of the future may derive valuable and fertile methods....*
>
> *From the straight line of Euclid to the lines of force of Faraday this has been the character of the ideas by which science has been advanced, and by the free use of dynamical as well as geometrical ideas we may hope for further advance.*"

Maxwell concluded his tribute to Faraday with the following paragraph:

> "*We are probably ignorant even of the name of the science which will be developed out of the materials we are now collecting, when the great philosopher next after Faraday makes his appearance.*"

Today, more than 135 years later, it is perhaps appropriate to repeat these words to conclude our sketch of the evolution of the physicists' description of the fundamental forces of nature.

Acknowledgments

We are indebted to Blaine Lawson for discussions about the evolution of ideas in mathematics.

It is our pleasure to thank Mrs. Shien-Ming Wu, the founding Director of the Shapiro Science Library of the University of Michigan, for her courtesy and for the excellent service we received from the library staff, especially in the retrieval of many old books and journals from the library resources without which our present study would have been much more limited in coverage.

We thank Ms. Judy Wong at the Chinese University of Hong Kong for her most efficient and capable assistance during the revisions of the manuscript.

Appendix. Historical Notes on Stokes' Theorem

This theorem was first stated in a letter, dated July 2, 1850, from Thomson to Stokes. Stokes acknowledged on July 5, 1850, that it was elegant and new. Stokes used this as question No. 8 on the Smith Prize examination in 1854 (the year Maxwell took it), which has since come to be known as the Stokes' Theorem. In 1864, Maxwell used this theorem without attribution. The theorem was first published in Thomson–Tait's *Treatise of Natural Philosophy*.[117] It appeared in Thomson's paper, "On Vortex Motion."[118] In 1873, the theorem appeared in Maxwell's *Treatise*, Vol. 1, Sec. 24. Prior to that, Maxwell wrote to Stokes on January 11, 1871, and to Tait on April 4, 1871, inquiring about the history of this theorem. Stokes' reply is lost, but Tait's reply credited this theorem to Thomson. For an interesting historical account of the Stokes' Theorem, the reader is referred to the article by J. J. Cross[10] and subsequently published volumes of the correspondence between Stokes–Thomson,[135] and *The Scientific Letters and Papers of J. C. Maxwell*[31–33] and references therein. A summary account is given below.

July 2, 1850

Letter from Thomson to Stokes (1819–1903) (Ref. 135, Vol. 1, p. 63):

My Dear Stokes

"... Do you know that the condition that $\alpha\,dx + \beta\,dy + \gamma\,dz$ may be the diff^{l} of a function of two indep^{t} variables for all points of a surface is

$$\lambda\left(\frac{d\beta}{dz} - \frac{d\gamma}{dy}\right) + m\left(\frac{d\gamma}{dx} - \frac{d\alpha}{dz}\right) + n\left(\frac{d\alpha}{dy} - \frac{d\beta}{dx}\right) = 0\,?$$

I made this out some weeks ago with ref^{ce} to electromagnetism. With ref^{ce} to an elastic solid, the cond^{n} may be expressed thus — the resultant axis of rotation at any point of the surface must be perp^{r} to the normal.

Yours very truly
William Thomson

P.S. The following is also interesting, and is of importance with reference to both physical subjects.

$$\int (\alpha\,dx + \beta\,dy + \gamma\,dz)$$
$$= \pm \iint \left\{\lambda\left(\frac{d\beta}{dz} - \frac{d\gamma}{dy}\right) + m\left(\frac{d\gamma}{dx} - \frac{d\alpha}{dz}\right) + n\left(\frac{d\alpha}{dy} - \frac{d\beta}{dx}\right)\right\} dS\,,$$

where λ, m, n denote the dir^{n} cosines of a normal through any el^{t} dS of a [limited] surface; and the integ^{n} in the sec^{d} member is performed over a portion of this surface bounded by a curve round w^{h} the int^{n} in the 1st member is performed.

July 5, 1850

Stokes' prompt reply to Thomson reads (Ref. 135, Vol. 1, p. 64):

Dear Thomson,

"... The theorems which you communicated are very elegant and are new to me. I have demonstrated them for myself, the first by the calculus of variations, the second (which includes the first as a particular case) by simple considerations like what I have employed at the beginning of my paper. I suppose from what you say that you arrived at these theorems by working with magnetic ideas....

Yours very truly
G. G. Stokes

1854: Stokes used this as Question No. 8 in the February 1854 Smith's Prize Examination, the year Maxwell took it.

1864: Maxwell used this theorem in his paper without attribution.

1867: The theorem was first published in Thomson–Tait's *Treatise of Natural Philosophy* (Ref. 117, article 190j).

1869: It appeared in Thomson's paper, Ref. 118. There Thomson states that "In a purely analytical light the result has an important bearing on the theory of the integration of complete or incomplete differentials. It was first given, with the indication of a more analytical proof than the preceding, in Thomson and Tait's *Natural Philosophy*, Sec. 190(j)."

1871: January 11, Maxwell wrote to Stokes: (Ref. 32, p. 351)

My dear Stokes

...Did not you set the theorem about the surface integral

$$\left(\left(\frac{d\beta}{dz}-\frac{d\gamma}{dy}\right)dy\,dz+\cdots+\right)$$

over the surface bounded by the curve s being equal to

$$\left(\frac{\alpha dx}{ds}+\frac{\beta dy}{ds}+\frac{\gamma dz}{ds}\right)ds\,.$$

I have had some difficulty in tracing the history of this theorem. Can you tell me something about it.

Yours truly
J. Clerk Maxwell

I hope you saw the eclipse well.

Unfortunately, Stokes' reply to Maxwell's inquiry has not been preserved. However, on April 4, 1871, Maxwell wrote to Tait on a postcard (Ref. 32, p. 366), asking the same question:

Dr. T′. ...But the history of

$$\iint\left\{\lambda\left(\frac{dZ}{dy}-\frac{dY}{dz}\right)+m\left(\frac{dX}{dz}-\frac{dZ}{dx}\right)+n\left(\frac{dY}{dx}-\frac{dX}{dy}\right)\right\}dS$$
$$=\int\left(\frac{Xdx}{ds}+\frac{Ydy}{ds}+\frac{Zdz}{ds}\right)ds$$

ascends (at least) to Stokes Smiths Prize paper 1854 and it then not altogether new to yours truly. Do you know its previous history? Poisson? on light???....

Yrs dp/dt

(Note added: T′ is the nickname of Tait, and dp/dt is the nickname of Maxwell in reference to a statement in thermodynamics: $dp/dt =$ JCM, the initial of Maxwell.)

Tait's reply to Maxwell is fortunately available (*Manuscripts in the University Library*, Cambridge, cited by Cross and as footnote 5 to Ref. 32, p. 366 by Harman)

> O $dp/dt, \ldots$ As to $\iint SUv\nabla\sigma\, ds = \int S\sigma\, d\rho$ I really thought it due to T and first published by the Archiepiscopal pair....
>
> T′

(Note added: Here Tait wrote the integral theorem in the quaternion form. T is the nickname of Thomson. The "Archiepiscopal pair" refers to Thomson and Tait — with the Archbishops of York and Canterbury having the same surnames. See, footnote 5 of Ref. 32 cited above.)

1873: In Maxwell's *Treatise*, Vol. 1, Sec. 24, the theorem is stated and proved. A footnote reads:

> "This theorem was give by Professor Stokes, *Smith's Prize Examination*, 1854, Question No. 8. It is proved in Thomson and Tait's *Natural Philosophy*, Sec. 190(j)."

1888: The 1888 edition of Thomson and Tait's *Natural Philosophy*, Sec. 190(j) contains the following footnote:

> "This theorem was given by Stokes in his Smith's Prize paper for 1854 (Cambridge University Calendar, 1854). The demonstration in the text is an expansion of that indicated in our first edition (1867). A more synthetic proof is given in Sec. 60(q) of Sir W. Thomson's paper on "*Vortex Motion*," *Trans. R.S.E.* 1869. A thoroughly analytical proof is given by Prof. Clerk Maxwell in his *Electricity and Magnetism* (Sec. 24)." (1873, note added)

1905: Some of the *Math. Tripos Examinations* and the *Smith's Prize Examinations* at Cambridge University were reprinted in the Appendix of Vol. 5 (1905) of the *Mathematical and Physical Papers of G. G. Stokes* (1819–1903), edited by J. Larmor, who added the following footnote to Question No. 8 of 1854:

> "This fundamental result, traced by Maxwell (*Electricity*, Vol. 1, Sec. 24) to the present source, has of late years been known universally as Stokes' theorem. The same kind of analysis had been developed previously in particular cases in Ampère's memoirs on electrodynamics of linear electric currents. And in a letter from Lord Kelvin, of date July 2, 1850, relating to such transformations, which has been found among Stokes' correspondence, the theorem in the text is in fact explicitly stated as a postscript. The vector which occurs in the surface integral had been employed by MacCullagh, who recognized its invariance, about 1837, in optical dynamics. It reappeared in Stokes' hands in 1845 (*ante*, Vol. I, p. 81) as twice the differential rotation in the theory of fluid motion and formed the basis of the mathematical

theory of viscosity of fluids, as also at a later time of Helmholtz's theory of vortex motion; its application to vibrations of solid elastic media was developed by Stokes in 1849 (*ante*, Vol. II, p. 253). Thus the theorem, though first stated by Lord Kelvin, relates to a quantity which, as regards physical applications, may be claimed to be Stokes' vector."

References

1. Y. Aharonov and D. Bohm, *Phys. Rev.* **115**, 485 (1959).
2. A. M. Ampère, *Mémoires de l'Académie Royale des Sciences* **6**, 175 (1827) [see especially pp. 252 and 253].
3. A. M. Ampère, *Theorie Mathematique des Phénomènes Électrodynamiques Uniquement Déduite de l'Expérience*, 2nd edn. (Hermann, Paris, 1883) [1st edn. published in 1826; to which we have no access].
4. V. I. Arnold, *Huygens and Barrow, Newton and Hooke: Pioneers in Mathematical Analysis and Catastrophe Theory from Evolvents to Quasicrystals*, translated from Russian by E. J. F. Primrose (Birkhäuser, Boston, 1990), p. 107.
5. M. F. Atiyah, N. J. Hitchin and I. M. Singer, *Proc. R. Soc. London, Ser. A* **362**, 425 (1978).
6. A. M. Bork, *Science* **152**, 597 (1966).
7. M. Born, *Vorlesungen über Atommechanik* (1924) [English translation: *The Mechanics of Atom* (Bell & Son, London, 1927), p. 208].
8. J. Z. Buchwald, *From Maxwell to Microphysics: Aspects of Electromagnetic Theory in the Last Quarter of the Nineteen Century* (University of Chicago Press, 1985).
9. R. G. Chambers, *Phys. Rev. Lett.* **5**, 3 (1960).
10. J. J. Cross, Integral theorems in Cambridge mathematical physics, 1800-55, in *Wranglers and Physicists*, ed. P. M. Harman (Manchester University Press, 1985), pp. 112–148 [see especially pp. 118–121].
11. O. Darrigol, *Electrodynamics from Ampère to Einstein* (Oxford University Press, 2000).
12. P. Debye, *Physik. Zeitschr.* **17**, 507 (1916).
13. B. S. DeWitt, *Dynamical Theory of Groups & Fields* (Gordon & Breach, New York, 1965).
14. J. Dieudonné, *A History of Algebraic and Differential Topology 1900–1960* (Birkhäuser, 1989).
15. P. A. M. Dirac, *Proc. Camb. Phil. Soc.* **23**, 69 (1926).
16. P. A. M. Dirac, *Proc. R. Soc. London, Ser. A* **133**, 60 (1931).
17. C. Ehresmann, Les connexions infinitesimals dans un espace fibre differentiable, in *Colloque de Topologie* (Bruxelles, 1950), pp. 29–55.
18. C. P. Enz, *No Time to be Brief: A Scientific Biography of Wolfgang Pauli* (Oxford University Press, 2002).
19. M. Faraday, *Experimental Researches in Electricity*, Vols. 1–3 (London, 1839–1855) [see Ref. 79].
20. M. Faraday, *Phil. Trans. R. Soc. London* **142**, 25 (1852).
21. M. Faraday, *Phil. Mag., Ser. 4* **3**, 401 (1852) [M. Faraday, *Electricity*, Vol. 3 (1852), pp. 328–370, 407–437].
22. V. Fock, *Zeit. für Physik* **39**, 226 (1926).
23. G. Green, *An Essay on the Application of Mathematical Analysis to the Theories of Electricity & Magnetism*, ed. T. Wheelhouse (Nottingham, 1828) [A facsimile

edition was published by U. Nottingham to celebrate Green's bicentennial in 1993]; see also, the *Mathematical Papers of the late George Green, Hermann* (Paris, 1903).

24. W. R. Hamilton, *Trans. R. Irish Acad.* **15**, 69 (1828).
25. W. R. Hamilton, *Trans. R. Irish Acad.* **16**, 1 (1830).
26. W. R. Hamilton, *Dublin Univ. Rev. & Quarterly Mag.* **1**, 795 (1833).
27. W. R. Hamilton, On the application to dynamics of a general mathematical method previously applied to optics, *British Assoc. Report* (Edinburgh, 1834), pp. 513–518.
28. W. R. Hamilton, On a general method in dynamics, *Phil. Trans. R. Soc.* **Part II**, 247–308 (1834).
29. W. R. Hamilton, Second essay on a general method in dynamics, *Phil. Trans. R. Soc.* **Part I**, 95–144 (1835).
30. P. M. Harman (ed.), Studies on Cambridge physics in 19th century, in *Wranglers and Physicists* (Manchester University Press, 1985).
31. P. M. Harman (ed.), *The Scientific Letters and Papers of James Clerk Maxwell, Vol. 1 (1846–62)* (Cambridge University Press, 1990).
32. P. M. Harman (ed.), *The Scientific Letters and Papers of James Clerk Maxwell, Vol. 2 (1862–73)* (Cambridge University Press, 1995).
33. P. M. Harman (ed.), *The Scientific Letters and Papers of James Clerk Maxwell, Vol. 3 (1874–79)* (Cambridge University Press, 2002).
34. O. Heaviside, *Electromagnetic Induction and its Propagation*, The Electrician (1885, 1886, 1887) [Ref. 36, Vol. 1, pp. 429–560].
35. O. Heaviside, *Phil. Mag.* (Jan. 1889) [Ref. 36, Vol. 2, pp. 468–485].
36. O. Heaviside, *Electrical Papers*, Vols. 1 & 2, The Electrician (London, 1892).
37. O. Heaviside, *Electromagnetic Theory*, Vol. 1, The Electrician (London, 1894).
38. O. Heaviside, *Electromagnetic Theory*, Vol. 2, The Electrician (London, 1899).
39. O. Heaviside, *Electromagnetic Theory*, Vol. 3, The Electrician (London, 1912).
40. H. von Helmholtz, "Über die Erhaltung der Kraft," Vorgetragen in der Sitzung der physikalische Gesellschaft zu Berlin am 23 Juli 1847 [Ref. 49, pp. 12–75].
41. H. von Helmholtz, *Selected Writings of H. von Helmholtz*, ed. K. Russell (Wesleyan University Press, 1971), pp. 3–55.
42. H. von Helmholtz, *Scientific Memoirs, Natural Philosophy*, eds. J. Tyndall and W. Francis (London, 1853), pp. 114–162.
43. H. von Helmholtz, *J. für Reine und Angew. Math.* **72**, 57 (1870).
44. H. von Helmholtz, Über die Theorie der Elektrodynamik, *Monatsberichte der Berliner Akademie* (1872), pp. 247–256 [Ref. 49, pp. 636–646].
45. H. von Helmholtz, Über die physikalische Bedeutung des Princips der kleinsten Wirkung, in *Borchardt-Crelle, Jour. für Mathematik*, Vol. 100 (1886), pp. 137–166, pp. 213–222 [Ref. 49, pp. 203–248].
46. H. von Helmholtz, Zur Geschichte des Princips der kleinsten Action, in *Sitzungsberichten der Akademie der Wissenschaften zu Berlin* (1887), pp. 225–236 [Ref. 49, pp. 249–263].
47. H. Helmholtz, Das princip der kleisten Wirkung in der Elektrodynamik, *Sitzungsber. Der Berliner Akad.* (May 12 1892).
48. H. von Helmholtz, *Wied. Ann.* **47**, 1 (1892) [Ref. 49, pp. 476–504].
49. H. von Helmholtz, *Wissenschaftliche Abhandlungen*, Vol. 1 (Leipzig, 1882).
50. H. von Helmholtz, *Wissenschaftliche Abhandlungen*, Vol. 3 (Leipzig, 1895).
51. H. R. Hertz, *Wiedemann's Ann.* **23**, 84 (1884) [Ref. 59, Article No. 17, pp. 273–290].
52. H. R. Hertz, *Gottinger Nachr.* **19** (1890).
53. H. R. Hertz, *Wiedemann's Ann.* **40**, 577 (1890) [Ref. 55, Article No. 13, pp. 208–255].
54. H. R. Hertz, *Wiedemanns Ann.* **41**, 369 (1890) [Ref. 55, Article No. 14, pp. 256–285].

55. H. R. Hertz, Unterschungen über die Ausbreitung der Electrischen Kraft (J. A. Barth, Leipzig, 1892).
56. H. R. Hertz, *Gesammelte Werke*, Vol. 2 (Leipzig, 1894), being 2nd edn. of Ref. 55.
57. H. R. Hertz, *Schriften vermischten Inhalts, Gesammelte Werke*, Vol. 1 (Leipzig, 1895).
58. H. R. Hertz, *Die Prinzipien der Mechanik, Gesammelte Werke*, Vol. 3 (Leipzig, 1894).
59. H. R. Hertz, *Miscellaneous Papers* [English translation: D. E. Jones and G. A. Schott (MacMillan, London, 1896)].
60. B. J. Hunt, *The Maxwellians* (Cornell University Press, 1991).
61. J. D. Jackson and L. B. Okun, *Rev. Mod. Phys.* **73**, 663 (2001).
62. H. B. Jones, *The Life & Letters of Faraday*, Vols. 1 and 2 (London, 1870).
63. C. Jungnickel and R. McCormmach, *Intellectual Mastery of Nature: Theoretical Physics from Ohm to Einstein, Vol. 1: The Torch of Mathematics 1800–1870* (University Chicago Press, 1986).
64. C. Jungnickel and R. McCormmach, *Intellectual Mastery of Nature: Theoretical Physics from Ohm to Einstein, Vol. 2: The Now Mighty Theoretical Physics 1870–1925* (University Chicago Press, 1986).
65. G. R. Kirchhoff, *Ann. der Physik* **100**, 193, 251 (1857) [G. R. Kirchhoff, *Gesammelte Abhandlungen* (Leipzig, 1882), p. 131].
66. G. R. Kirchhoff, *Ann. der Physik* **102**, 529 (1857) [G. R. Kirchhoff, *Gesammelte Abhandlungen* (Leipzig, 1882), p. 154].
67. P. S. Laplace, *Traité de Mécanique Céleste*, Vols. 1–5 (Duprat, Paris, 1825) [Vol. 1, pp. 136–137; See also, *Öuvres Complète de Laplace* (14 volumes) (Gauthier-Villars, 1843–1912)].
68. J. Larmor, *Proc. London Math. Soc.* **15**, 158 (1894) [J. Larmor, **1**, pp. 31–70 (1929)].
69. J. Larmor, *Phil. Trans. R. Soc.* **185**, 719 (1894) [Ref. 73, pp. 414–535, especially p. 524].
70. J. Larmor, *Phil. Trans. R. Soc.* **186**, 695 (1895) [Ref. 73, pp. 543–597, especially p. 568].
71. J. Larmor, *Aether and Matter* (Cambridge University Press, 1900) [especially p. 94].
72. J. Larmor, *Trans. Camb. Phil. Soc.* **18**, 380 (1900) [Ref. 74, pp. 158–191, especially p. 166].
73. J. Larmor, *Mathematical and Physical Papers*, Vol. 1 (Cambridge University Press, 1929).
74. J. Larmor, *Mathematical and Physical Papers*, Vol. 2 (Cambridge University Press, 1929).
75. F. London, *Zeit. für Physik* **42**, 374 (1927).
76. H. A. Lorentz, *Versuch einer Theorie der Elektrischen und Optischen Erscheinungen in bewegt Korpen* (Leiden, 1895) [H. A. Lorentz, *Collected Papers*, Vol. 5, pp. 1–137 (The Hague, M. Nijoff, 1937)].
77. H. A. Lorentz, Maxwells elektromagnetische Theorie, in *Encyklopadie Math. Wissensch.* **5-2** (Leipzig, 1904), pp. 63–144.
78. H. A. Lorentz, Weiterbildung der maxwellschen Theorie. elektronentheorie, in *Encyklopadie Math. Wissensch.* **5-2** (Leipzig, 1904), pp. 145–280.
79. T. Martin (ed.), *Faraday's Diary (1820–1862)*, Vols. 1–7, plus index (G. Bells & Sons, London, 1932–1936).
80. J. C. Maxwell, *Trans. Camb. Phil. Soc.* **10**, 27 (1856) [Ref. 95, Vol. 1, pp. 155–229, referred to as paper 1].
81. J. C. Maxwell, *Phil. Mag.* **21**, 161 (1861), as Part I of paper 2 [Ref. 95, Vol. 1, pp. 451–466].

82. J. C. Maxwell, *Phil. Mag.* **21**, 281 (1861), as Part II of paper 2.
83. J. C. Maxwell, *Phil. Mag.* **21**, 338 (1861), as Part II of paper 2 [Ref. 95, Vol. 1, pp. 467–488].
84. J. C. Maxwell, *Phil. Mag.* **23**, 12 (1862), as Part III of paper 2.
85. J. C. Maxwell, *Phil. Mag.* **23**, 85 (1862), as Part IV of paper 2 [Ref. 95, Vol. 1, pp. 489–513].
86. J. C. Maxwell, *Phil. Trans. R. Soc. London* **155**, 459 (1865) [Ref. 95, Vol. 1, pp. 526–571, referred to as paper 3].
87. J. C. Maxwell, *Phil. Trans.* **158**, 643 (1868) [Ref. 95, Vol. 2, pp. 125–143].
88. J. C. Maxwell, *A Treatise on Electricity and Magnetism*, Vols. 1 and 2 (Oxford, 1873).
89. J. C. Maxwell, *A Treatise on Electricity and Magnetism*, Vols. 1 and 2, 2nd edn., ed. W. D. Niven (Oxford, 1881).
90. J. C. Maxwell, *A Treatise on Electricity and Magnetism*, Vols. 1 and 2, 3nd edn., ed. J. J. Thomson (Oxford, 1892).
91. F. E. Neumann, *Abhandlungen der Berliner Akademie der Wissenschaften* **33** (1845) [F. E. Neumann, *Gesammelte Werke*, Vol. 3, eds. C. Neumann *et al.* (Leipzig, 1912), pp. 259–343].
92. F. E. Neumann, *Abhandlungen der Berliner Akademie der Wissenschaften* **44** (1847) [F. E. Neumann, *Gesammelte Werke*, Vol. 3, eds. C. Neumann *et al.* (Leipzig, 1912), pp. 347–424].
93. F. E. Neumann, *Vorlesungen über die Theorie des Magnetismus* (*Magnetischen Induktion*) (*1857*), Mathematische Physik, Vol. 1 (Leipzig, 1881).
94. F. E. Neumann, *Vorlesungen über elektrische Ströme* (*1864–1865*), Mathematische Physik, Vol. 3 (Leipzig, 1884).
95. W. D. Niven (ed.), *The Scientific Papers of James Clerk Maxwell*, Vols. 1 and 2 (Cambridge University Press, 1890).
96. L. O'Raifeartaigh, *The Dawning of Gauge Theory* (Princeton University Press, 1997).
97. L. O'Raifeartaigh and N. Straumann, *Rev. Mod. Phys.* **72**, 1 (2000).
98. W. Pauli, Die Allgemeinen Prinzipien der Wellenmechanik, *Handbuchs der Physik*, Vol. 34 (1933) [S. Flugge (ed.), Springer Series, Vol. 5, Part I (1958)].
99. W. Pauli, *Rev. Mod. Phys.* **13**, 203 (1941).
100. J. Renn and R. Schulmann, *Albert Einstein/Mileva Maric*: *The Love Letters* (Princeton University Press, Princeton, New Jersey, 1992).
101. G. A. Schott, *Electromagnetic Radiation* (Cambridge University Press, Cambridge, 1912), especially App. F, pp. 282–296.
102. K. Schwarzschild, *Gött. Nachr. Math.-Phys. Kl.* 126 (1903).
103. K. Schwarzschild, *Gött. Nachr. Math.-Phys. Kl.* 132 (1903).
104. K. Schwarzschild, *Gött. Nachr. Math.-Phys. Kl.* 245 (1903).
105. I. M. Singer, Some problems in the quantization of gauge theories and string theories, in *Proc. Symposia in Pure Math.*, Vol. 48 (1988), pp. 198–216.
106. A. Sommerfeld, *Physik Zeitschr.* **17**, 491 (1916).
107. M. Spivak, *A Comprehensive Introduction to Differential Geometry*, 2nd edn. (Publish or Perish, Inc., Berkeley, CA, 1979).
108. G. G. Stokes, *Trans. Camb. Phil. Soc.* **8**, 287 (1845) [Ref. 110, pp. 75–129].
109. G. G. Stokes, *Trans. Camb. Phil. Soc.* **9**, 1 (1849), especially pp. 9 and 13 [Ref. 111, pp. 243–328].
110. G. G. Stokes, *Mathematical and Physical Papers*, Vol. 1 (Cambridge University Press, 1880).

111. G. G. Stokes, *Mathematical and Physical Papers*, Vol. 2 (Cambridge University Press, 1883).
112. G. G. Stokes, *Mathematical and Physical Papers*, Vol. 5 (Cambridge University Press, 1905).
113. W. Thomson, *Camb. & Dublin Math. J.* **2** (1847) [W. Thomson, *Mathematical & Physical Papers*, Vol. 1 (Cambridge University Press, 1882), pp. 76–80].
114. W. Thomson, *Phil. Trans. R. Soc. London* **141**, 243 (1851).
115. W. Thomson, *Phil. Trans. R. Soc. London* **141**, 269 (1851) [W. Thomson, Ref. 119, Paper No. 24, pp. 340–404; especially Chap. V: "On Solenoidal and Lamellar Distributions of Magnetism," article 521, p. 402 (1872)].
116. W. Thomson, *Phil. Mag. Ser. 4* **1**, 177 (1851) [Ref. 119, pp. 465–480].
117. W. Thomson and P. G. Tait, *Treatise on Natural Philosophy*, Vol. 1, Part 1 (Oxford, 1867), we have access only to the 1879 and 1888 editions.
118. W. Thomson, *Trans. R. Soc. Edinburgh* **25** (1869), Sec. 60(q).
119. W. Thomson, Reprints of Papers on *Electrostatics & Magnetism* (Macmillan, London, 1872).
120. A. Tonomura *et al.*, *Phys. Rev. Lett.* **48**, 1443 (1982).
121. A. Tonomura *et al.*, *Phys. Rev. Lett.* **56**, 792 (1986).
122. J. Tyndall, *Faraday as a Discoverer* (D. Appleton, New York, 1873).
123. J. H. Van Vleck, *Bull. Nat. Res. Council* **10** (1926), Part 4, No. 54.
124. W. E. Weber, Elektrodymische Maassbestimungen, über ein allgemeines Grundgesetz der elektrischen Wirkung, *Abhandlungen bei Begrund der Koniglichen Sachsischen Gesellschaft der Wissenschaften* (Leipzig, 1846), pp. 211–378 [W. E. Weber, *Wilhelm Weber's Werkes*, Vols. 1–6 (Berlin, 1892–1894); Vol. 3, pp. 25–214].
125. H. Weyl, *Math. Zeit.* **2**, 384 (1918) [Ref. 131, Vol. 2, pp. 1–28].
126. H. Weyl, *Sitzbe. Preuss Akad. Wiss.* 465 (1918) [Ref. 131, Vol. 2, p. 465; English translations: *Gravitation and Electricity*, appeared in (a) *The Principle of Relativity* (A collection of original memoirs on the special and general relativity by H. A. Lorentz, A. Einstein, H. Minkowski and H. Weyl; with notes by A. Sommerfeld) (Methusen & Co., Ltd., London 1923), transl. by W. Perrett and G. B. Jeffrey from the German text *Das Relativitätsprincizip* (Teubner, Leipzig, 1922), reprinted as Dover Publications in 1952; (b) in Ref. 96].
127. H. Weyl, *Ann. der Physik* **59**, 101 (1919) [Ref. 131, Vol. 2, p. 55. The above three papers will be referred to as Weyl 1, 2, 3].
128. H. Weyl, Gravitation and the electron, *Proc. U.S. Nat. Acad. Sci.* **15**, 323 (1929) [Ref. 131, Vol. 3, p. 217 (translation into English from the German manuscript by H. P. Robertson). W. Pauli was said to be critical of this paper of Weyl, but was later mollified after reading Weyl's publication in 1929 (Ref. 130), see the publications by O'Raifeartaigh in 1997 (Ref. 96) and Enz in 2002 (Ref. 18)].
129. W. Weyl, *Gravitation and the Electron*, Lectures given in May 20–23, 1929 at Rice University, appeared as The Rice University Studies, Vol. 16, No. 4, pp. 280–295, in Ref. 131, Vol. 3, p. 229.
130. H. Weyl, *Zeit. für Physik* **56**, 330 (1929) [Ref. 131, Vol. 3, p. 245. An English translation appeared in O'Raifeartaigh Ref. 96 (1997)].
131. H. Weyl, *Gesammelte Abhandlungen*, Vols. 1–4, ed. K. Chandrasekharan (Springer-Verlag, Berlin, 1968).
132. E. T. Whittaker, *A History of the Theories of Aether and Electricity* (*From the Age of Descartes to the Close of the 19th Century*) (Longmans, Green & Co., London, 1910), p. 224.
133. E. T. Whittaker, in *Lives in Science* (Simon and Schuster, 1957).

134. L. P. Williams, *Michael Faraday: A Biography* (Basic Books, New York, 1965).
135. D. D. Wilson, The Correspondence between Sir George Gabriel Stokes and Sir William Thomson, Baron Kelvin of Largs (Cambridge University Press, 1990) [Vol. 1, pp. 1–342 (1846–1869); Vol. 2, pp. 343–783 (1870–1901)].
136. T. T. Wu and C. N. Yang, *Phys. Rev. D* **12**, 3845 (1975).
137. T. T. Wu and C. N. Yang, *Phys. Rev. D* **14**, 437 (1976).
138. C. N. Yang and R. Mills, *Phys. Rev.* **95**, 631 (1954).
139. C. N. Yang and R. Mills, *Phys. Rev.* **96**, 191 (1954).
140. C. N. Yang, *Phys. Rev. Lett.* **33**, 445 (1974).
141. C. N. Yang, *Annals of NY Academy of Sciences* **294**, 86 (1977).
142. C. N. Yang, *Phys. Today* **33**, 42 (1980).
143. C. N. Yang, *Selected Papers 1945–1980 with Commentary* (Freeman, 1983).
144. C. N. Yang, Gauge fields, electromagnetism and the Bohm–Aharonov effect, in *Proc. Int. Symp. Foundation of Quantum Mechanics*, Tokyo, 1983, eds. S. Kamefuchi *et al.* (Physical Society of Japan, 1984), pp. 5–9.
145. C. N. Yang, Hermann Weyl's contribution to physics, in *Hermann Weyl 1885–1985*, ed. K. Chandrasekharan (Springer-Verlag, 1986), pp. 7–21.
146. C. N. Yang, Square root of minus one, complex phase and Erwin Schrödinger, *Schrödinger*, ed. C. W. Kilmister (Cambridge University Press, 1987), pp. 53–64.
147. C. N. Yang, Symmetry and physics (1988), in *The Oskar Klein Memorial Lectures*, Vol. 1, ed. G. Ekspong (World Scientific, 1991), pp. 11–33.
148. C. N. Yang, Maxwell's equations, vector potential and connections on fiber bundles, in *Proc. 3rd Int. Symp. Foundation of Quantum Mechanics*, Tokyo, 1989, eds. S. Kobayashi *et al.* (Physical Society of Japan, 1990), pp. 383–384.
149. C. N. Yang, Reflections on the development of theoretical physics, in *Proc. 4th Int. Symp. Foundation of Quantum Mechanics*, Tokyo, 1992, JJAP Series 9 (1993), pp. 3–9.
150. C. N. Yang, *Vector Potential, Gauge Field and Connection on a Fiber Bundle*, Vol. 3 (Tsinghua Science and Technology, 1997), pp. 861–870.

https://doi.org/10.1142/9789819826636_0019

BETHE'S HYPOTHESIS*

CHEN NING YANG[†] and MO-LIN GE[‡]
[†]*State University of New York, Stony Brook, New York and Tsinghua University, Beijing, China*
[‡]*Nankai Institute of Mathematics, Nankai University, Tianjin, China*
[†]*yang@castu.tsinghua.edu.cn*
[‡]*geml@nankai.edu.cn*

1. Early History

In 1931 Bethe solved,[1] rigorously, the spin wave collision problem of Bloch[2] in one dimension. He used an Ansatz about the algebraic structure of the wave function for the spin waves. This Ansatz has now blossomed into several active fields of research in physics and in mathematics

Before the Second World War the most important development along this line was in the work of Hulthen[3] in 1938. After the war, many papers used Bethe's hypothesis:[4] Orbach (1958), Walker (1959), Lieb, Schultz and Mattis (1961), des Cloizeaux and Pearson (1962), Katsura (1962), Lieb (1963), Lieb and Liniger (1963), Griffiths (1964), McQuire (1964) and others.

In 1966 C. N. Yang and C. P. Yang published three papers[5] on what is now called the XXZ model. They used Bethe's Ansatz and found it very powerful. So they thought it should be explicitly named to honor Bethe. Thus their first paper had the title:

One-Dimensional Chain of Anisotropic Spin-Spin Interactions
I. Proof of Bethe's Hypothesis for Ground State in a Finite System

That was where the name Bethe's Hypothesis had first appeared in print. Happily it has now taken hold as Bethe's Hypothesis, or Bethe's Ansatz, or simply BA.

We do not know how Bethe originally had arrived at his hypothesis. It is, in spirit, a generalization of the "only reflection" idea in the classical method of images, which is very old. In the mid-nineteenth century, "Electric Images" was a

*Also appeared in "Hans Bethe and His Physics", eds. G. E. Brown and C.-H. Lee, World Scientific, 2006.

Reproduced from International Journal of Modern Physics B, Vol. 20, No. 16, pp. 2223–2225 (2006), DOI: 10.1142/S0217979206034728

fashionable subject. In particular William Thomson was proud of his discovery[6] of images in spherical conductors. Maxwell in his *Treatise* wrote

> An electrical image is an electrified point or system of points on one side of a surface which would produce on the other side of that surface the same electrical action which the actual electrification of that surface really does produce.

In Bethe's hands, "only reflection" became generalized to "only reflection and transmission," i.e. "no diffraction." By including transmission the idea becomes more subtle, more complex and much more general, and that is the magic of Bethe's Hypothesis.

Could there be a case in 2 dimensions where the "no diffraction" hypothesis could apply? We have searched for such a case, but without success.

2. Later Developments

In retrospect, the key idea in Bethe's Hypothesis was in fact also latent in the star-triangle relations of statistical mechanics, and in Artin's knot theory. These developments led to the Yang-Baxter Equation (YBE). Many solvable models were

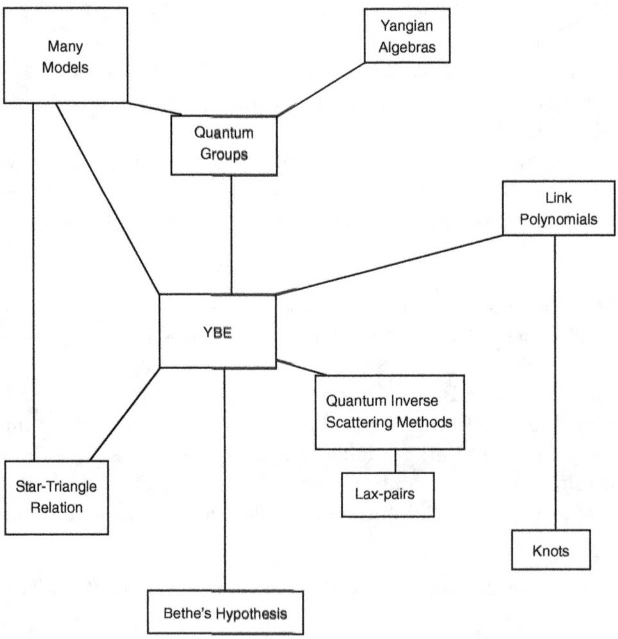

Fig. 1.

invented using ideas based on Bethe's Hypothesis and its generalizations throughout the 1970's. Then in the 1980's the idea of "quantum groups" led to explosive advances, especially by the Leningrad School, by the Japanese scientists, and by Drinfel'd. Around the same time Jones and others, starting from the algebraic side, made great advances in knot theory.

In Fig. 1 we present a schematic diagram of these developments in approximate time ordering. It is impossible to list all the key references in these intensive and extensive investigations in mathematics and in physics, so we shall not try to.

References

1. H. A. Bethe, *Z. Physik* **71**, 205 (1931).
2. F. Bloch, *Z. Physik* **61**, 206 (1930); **74**, 295 (1932).
3. L. Hulthen, *Arkiv Mat. Astron. Fysik* **26A**(1) (1938).
4. R. Orbach, *Phys. Rev.* **112**, 309 (1958); L. R. Walker, *Phys. Rev.* **116**, 1089 (1959); E. Lieb, T. Schultz and D. Mattis, *Ann. Phys. (N.Y.)* **16**, 407 (1961); J. des Cloizeaux and J. J. Pearson, *Phys. Rev.* **128**, 2131 (1962); S. Katsura, *Phys. Rev.* **127**, 1508 (1962); E. Lieb, *Phys. Rev.* **130**, 1616 (1963); E. Lieb and W. Liniger, *Phys. Rev.* **130**, 1605 (1963); R. B. Griffiths, *Phys. Rev.* **133**, A768 (1964); J. B. McQuire, *J. Math. Phys.* **5**, 622 (1964).
5. C. N. Yang and C. P. Yang, *Phys. Rev.* **150**, 321 (1966); *ibid.* **150**, 327 (1966); *ibid.* **151**, 258 (1966).
6. W. Thomson, *Cambridge and Dublin Math. J.* **5**, 1 (1850).

https://doi.org/10.1142/9789819826636_0020

SOME PROBLEMS IN COLD ATOM RESEARCH*

C. N. YANG
Tsinghua University, Beijing 100084, China

It is a great pleasure to be here today, seeing so many old friends to celebrate the 80th birthday of my longtime friend. I think, probably, of all the people in the audience, I am the one who have known him for the longest period. I met Murray in 1951 when I was at the Institute for Advanced Study in Princeton.

Listening to the talks this morning, I sort of re-experienced the long career I have in elementary particle physics, but I also work in statistical mechanics. We all know Murray has a wide range of interests but probably, few people in this audience know that he has also written good papers in statistical mechanics.

This (Fig. 1) is one of the papers ("Correlation Energy of an Electron Gas at High Density") he wrote with Brueckner — a very good paper. I think he left that year, afterwards, never to return.

In recent years, the field of cold atoms has become very, very hot. The reason is because of the new, incredible technology both in laser technology and microelectronics.

Here [Fig. 2(a)] is a chip manufactured in the laboratory in Amsterdam. The structure of the chip is illustrated here [Fig. 2(b)]. There is a current, I, in red, which generates a cylindrical magnetic field, and there is also a biased field going that way. The two of them could produce together a reading of a small magnetic field, or near-zero magnetic field, but with extremely strong gradient in the neighborhood. That became a one-dimensional trap suspended 90 microns above the chip. As a consequence, you can trap thousands of atoms, either fermions or bosons in that trap. That is becoming an extremely interesting new field of study.

The theory of it is actually very simple, in principle. What we need to study is the Hamiltonian,

$$H = \sum_{i=0}^{N} \left[-\frac{1}{2}\frac{\partial^2}{\partial x_i^2} + \frac{1}{2}x_i^2 \right] + g \sum_{i>j} \delta(x_i - x_j)\,, \quad g \text{ tunable}, \; e_n = \frac{1}{2} + n\,. \tag{1}$$

*Invited talk given at the conference in Honour of Murray Gell-Mann's 80th Birthday, 24–26 February 2010, Singapore.
Transcribed by C. H. Oh (National University of Singapore). Unread by C. N. Yang.

Reproduced from International Journal of Modern Physics B, Vol. 24, No. 18, pp. 3469–3477 (2010), DOI: 10.1142/S0217979210056104

PHYSICAL REVIEW VOLUME 106, NUMBER 2 APRIL 15, 1957

Correlation Energy of an Electron Gas at High Density*

MURRAY GELL-MANN, *Department of Physics, California Institute of Technology, Pasadena, California*

AND

KEITH A. BRUECKNER, *Department of Physics, University of Pennsylvania, Philadelphia, Pennsylvania*

(Received December 14, 1956)

The quantity ϵ_c is defined as the correlation energy per particle of an electron gas expressed in rydbergs. It is a function of the conventional dimensionless parameter r_s, where r_s^{-3} is proportional to the electron density. Here ϵ_c is computed for small values of r_s (high density) and found to be given by $\epsilon_c = A \ln r_s + C + O(r_s)$. The value of A is found to be 0.0622, a result that could be deduced from previous work of Wigner, Macke, and Pines. An exact formula for the constant C is given here for the first time; earlier workers had made only approximate calculations of C. Further, it is shown how the next correction in r_s can be computed. The method is based on summing the most highly divergent terms of the perturbation series under the integral sign to give a convergent result. The summation is performed by a technique similar to Feynman's methods in field theory.

Fig. 1.

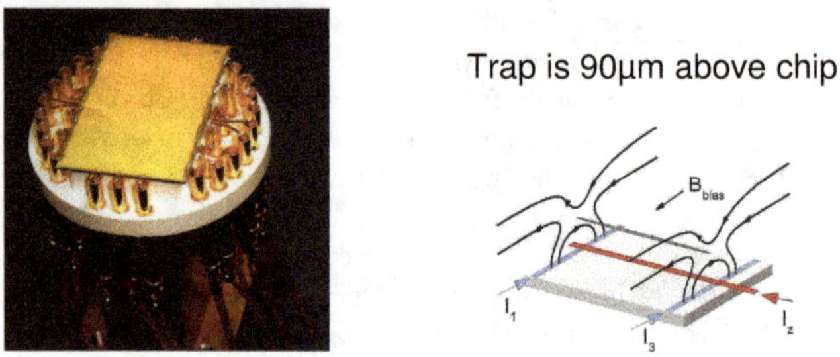

(a) (b)

Fig. 2.

The first part of the Hamiltonian in the square bracket is the harmonic trap potential, and the last part is a one-dimensional delta function interaction. You can easily prove that in one dimension, the interaction between two atoms is reducible to a term of this form. And in particular, the g can be tuned with the new laser magneto trap and technology. So, this became an extremely interesting theoretical problem.

If you do not have the g term, that is to say one-dimensional trap problem, which can be easily solved. If you do not have the trap, the harmonic trap, that problem is one-dimensional many-body, either bosons or fermions interact with the delta function potential, that problem had been solved in the 1960s by Lieb and by me. So, the present problem is a combination of the two — how do you deal with the trap and the delta function interaction. My colleague Ma and I had solved it and we had recently published it.

Let me explain to you what is the strategy. If you do not have the g term, it is just a one-dimensional harmonic oscillator and the energy of each particle is E_n, which is equal to $\frac{1}{2} + n$ in the right units [Eq. (1)].

Now, let us look at the ground state energy, which is E_0 for this system, for different values of g (Fig. 3). First, let us concentrate on this point, which means

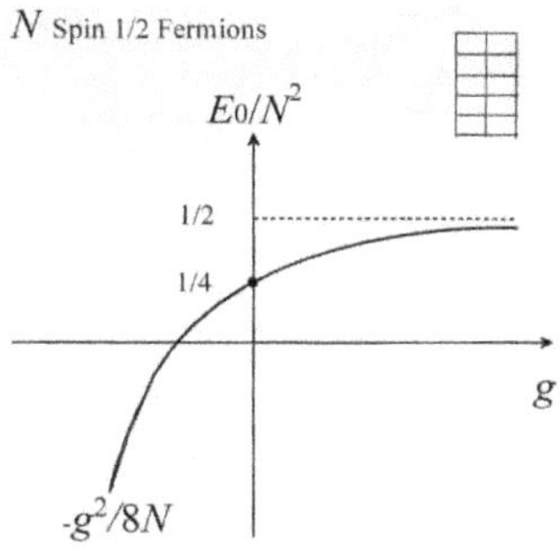

Fig. 3.

there is no delta interaction, this is a system of free fermions. Now for spin one-half fermion, you have various symmetries and it had been proven long time ago by Lieb, that the ground state has this symmetry of space wave function, that is the Young tableau. It is obvious that without the g term, it is just a collection of non-interacting fermions, and therefore for the ground state, you can easily find the energy. Obviously, half of the particles has spin-up and half of the particles with spin-down. Half of the particles with spin-up fill up the space with $n = 0$, $n = 1$, $n =$ so on and so forth. So clearly the total energy is given by $N^2/4$. So, we know that when E_0/N_2 for $g = 0$, is equal to $1/4$:

$$\boxed{g = 0} \quad \begin{matrix} N/2 \text{ spin} \uparrow \\ N/2 \text{ spin} \downarrow \end{matrix} \quad E_0 \cong 2\sum_{0}^{N/2}\left(\frac{1}{2} + n\right) \cong \frac{N^2}{4}. \tag{2}$$

Let us look at this graph (Fig. 3), that is for very large g, for N fermions. If you look at the large positive g for N fermions. These N fermions have two kinds of spins, some spin-up and some spin-down but the coupling constant is very large and is positive. That means the wave function has to be zero when the two particles are together. That is true of course if they have the same spin, but for $g = +\infty$, even if they have one spin-up and one spin-down, so the Pauli's exclusion principle does not make it equal to zero, but the delta function would make it equal to zero.

In short, when $g = \infty$, the N particles have to occupy states, free particle states, from $N = 0$, all the way up to N, now up to $N = 2$. Therefore, the total energy is given by this, that is equal to $N^2/2$:

$$\boxed{g = +\infty} \quad N \text{ Fermions} \quad E_0 \cong \sum_{0}^{N}\left(n + \frac{1}{2}\right) \cong \frac{N^2}{2}. \tag{3}$$

So, we now should know if we go to this large g limit, $E_0/N^2 = 1/2$, go to $g = 0$, this is equal to $1/4$, which is half of that (Fig. 3). How about here where g is negative?

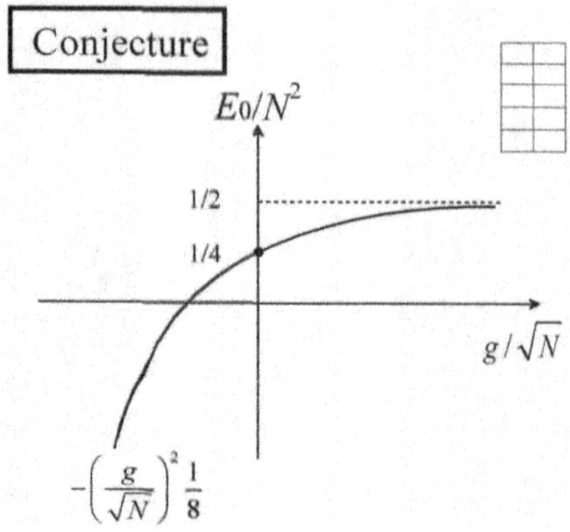

Fig. 4.

For g that is negative and large, the particle has the tendency to form spin up–spin down pairs with very deep attraction, and the pair energy is well-known. This was already done in the 1960s. The pair energy is equal to $-g^2/4$, and you have $N/2$ pairs. So, this is now the total energy:

$$\boxed{g = -\text{large}} \quad \text{pairs of } \uparrow\downarrow \text{ bound states} \quad E_0 \cong \left(\frac{N}{2}\right)\left(-\frac{g^2}{4}\right). \tag{4}$$

The total energy is proportional to Ng^2. So now, if you look at these (Fig. 3), it is very strange, because, here $g = \infty$, it is pretty constant. Here for negative g, there is a factor g^2/N. When Ma and I first looked at it, we realized what this indicates. You should plot it against $g/\sqrt{N}$ (Fig. 4), this is 1/4, and this is 1/2, this becomes this. This equals to $\sqrt{N}$, it is a function of this. In other words, we realize that in this plot, this becomes a limit when N goes to infinity. This was the conjecture that we published last year, and now we have proven that conjecture. In fact, we are now able to complete the whole curve.

The strategy of that is as follows. We later proved this conjecture using the Thomas–Fermi method. Thomas–Fermi method is well-known, of course, to every student of physics. It is a very physical idea.

Let us re-examine that. Let us first consider **free particles in harmonic trap with one component**, not spin-up and spin-down, just spin-up with N fermions. Let us consider this problem. Thomas–Fermi method said that in a harmonic trap like this, you consider a small region dx, and you treat the particle in here as if they are free, and you consider the particles within this interval. Then you add the total kinetic energy that would be from this point to here (Fig. 5). The maximum kinetic energy which is this length plus the potential should be equal to this level. You realize that for N fermions, the density at this point should be equal to this. Density as the function of x is a half ellipse (Fig. 6). So this is the Thomas–Fermi

Thomas-Fermi Method
(for one component Fermions)

$V = \frac{1}{2}x^2$

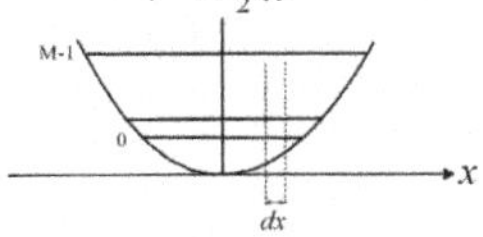

In interval dx

$$\text{Max K.E.} + V(x) = (M-1) + \frac{1}{2}$$

$$\text{Max K.E.} = \frac{1}{2}\left(\frac{\pi}{dx}n\right)^2 = \frac{1}{2}\pi\rho^2$$

$$\frac{1}{2}(\pi\rho)^2 + \frac{1}{2}x^2 \cong M$$

$$\rho(x) = \frac{1}{\pi}\sqrt{2M - x^2}$$

Fig. 5.

Limiting density as M→∞

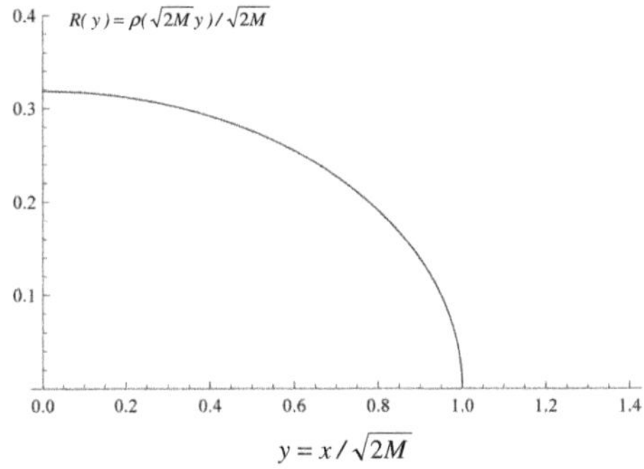

Fig. 6.

result of that problem. But of course, in this case, you can calculate exactly the density distribution because there is no coupling g:

$$\frac{\rho(x)}{\sqrt{2M}} = \frac{1}{\pi}\sqrt{1 - \left(\frac{x}{\sqrt{2M}}\right)^2},$$
$$\frac{x}{\sqrt{2M}} = y\,, \tag{5}$$
$$R(y) = \frac{\rho(\sqrt{2M}\,y)}{\sqrt{2M}} = \frac{1}{\pi}\sqrt{1-y^2}\,.$$

So, the density is just the sum of the M lowest state of particle density, these are the normalized wave functions:

$$\rho(x) = \sum_0^{M-1} |\psi_n(x)|^2\,. \tag{6}$$

If you plot that for $M = 5$, you get this expression (Fig. 7). Let us compare that with Thomas–Fermi, it is in very good agreement (Fig. 8). If we now do it for $M = 15$, the agreement is better (Figs. 9 and 10); if you do it for $M = 55$ particles, the agreement is excellent (Figs. 11 and 12). As the number of particles is

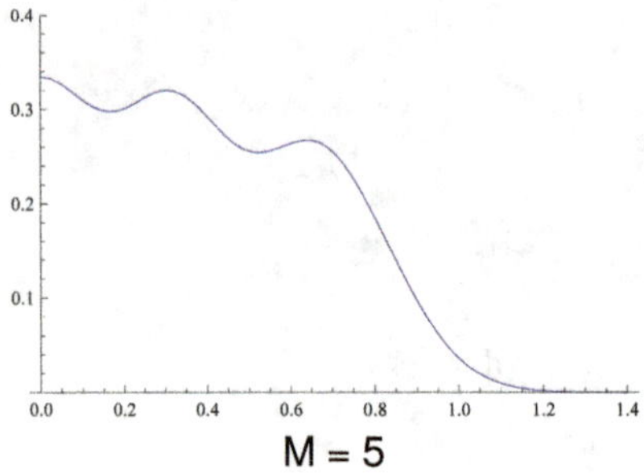

Fig. 7.

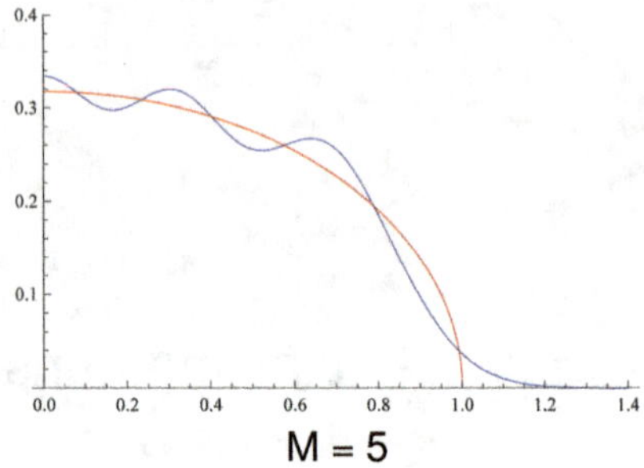

Fig. 8.

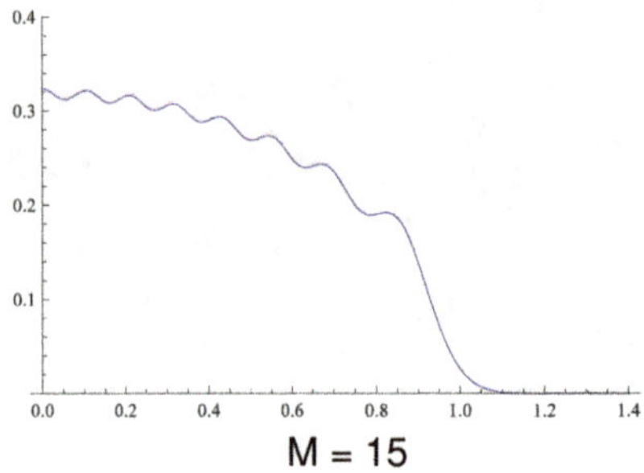

Fig. 9.

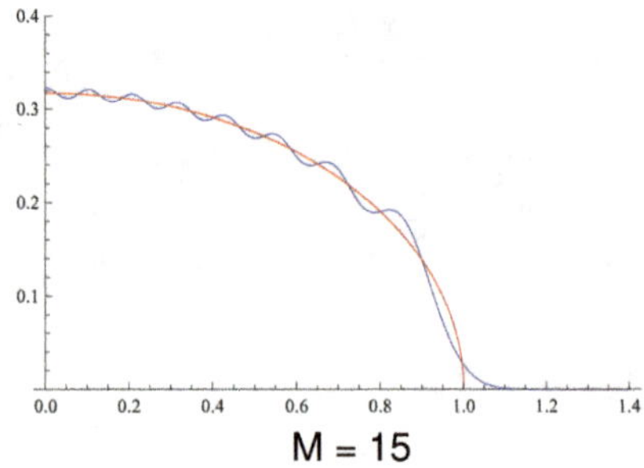

Fig. 10.

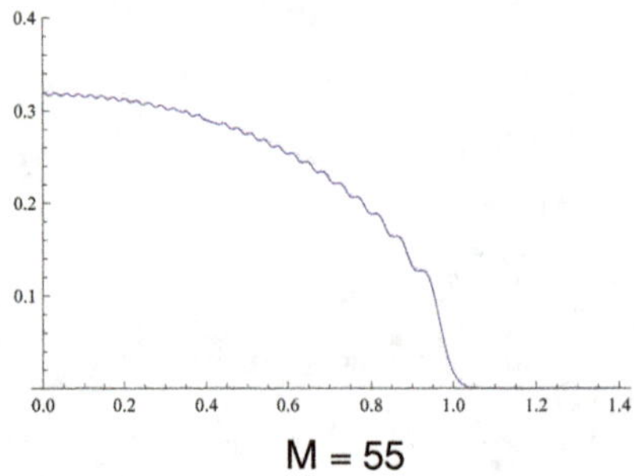

Fig. 11.

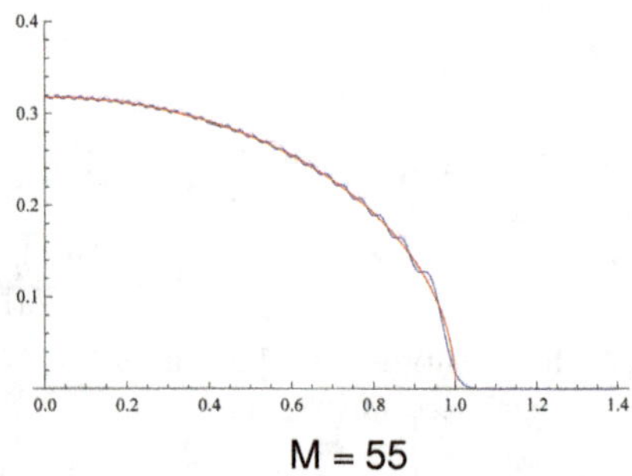

Fig. 12.

increased, the Thomas–Fermi method becomes the limit, in fact becomes the limit very rapidly.

So, we conclude that Thomas–Fermi method gives the limit, limiting density as M goes to infinity. Now we applied this idea to the case when you have both the trap and the g. But first, let us ask ourselves why the Thomas–Fermi method gives such a good limit.

As M goes to infinity, almost all particles have very small wavelengths. Therefore, they are classical particles. Therefore, the classical statistical mechanics holds. Therefore, the Thomas–Fermi method gives the correct limit. Once you appreciate this, you realize that the Thomas–Fermi method has a very wide-ranging applicability, in fact, not only in one-dimension, but in two dimensions and in three dimensions. This is the realization which I only reached within the last few months. At first, I thought that the good agreement that I showed you was because it was one-dimensional. We all know that two dimensions and one dimension are very different. Because in one dimension, there is no diffraction. That is a reason why Lieb and I were able to solve the one-dimensional problem without the trap using the Bethe ansatz. The spirit of Bethe ansatz is that there is no diffraction in one dimension. But now, I realize that is irrelevant as far as validity of the Thomas–Fermi method is concerned.

So, let us see how we do it. First, **without the harmonic potential** $V(x)$, **but with the delta function interaction** $\Sigma\delta(x_i - x_j)$, this problem was solved in 1963 by Lieb and Liniger for bosons, and by me in 1967 for fermions. We used the Bethe's hypothesis of 1931. In one dimension, if two particles p_1, p_2 collided, conservation of momentum gives the first equation [Eq. (7)], conservation of energy gives the second equation [Eq. (8)]:

$$p_1 + p_2 = p_1' + p_2' \,, \tag{7}$$

$$E_1 + E_2 = E_1' + E_2' \,. \tag{8}$$

But when we solved this, you find that there are only two solutions:

$$p_1' = p_1 \,, \quad p_2' = p_2 \,, \tag{9}$$

$$p_1' = p_2 \,, \quad p_2' = p_1 \,. \tag{10}$$

One is this equation [Eq. (9)] which means a straightforward scattering. This [Eq. (10)] means exchange. So, in one dimension, because of conservation of momentum and conservation of energy, you only have reflection and transmission, no diffraction, and that is the reason why Bethe ansatz works in one dimension, and that is also the reason why Bethe ansatz does not work in higher dimensions. Using this, we can solve the problem, in particular for the fermion case. This (slide) says that in N particles, many particles p_1, p_2, p_3, up to p_N, then in each condition, you have just the permutation without change, so all you can do is to permute these momenta, so the analytic problem, the differential equation problems become an algebraic problem. So that is a trick which makes the Bethe ansatz so powerful.

- Lieb & Linnoger 1963: Bosons, spinless.
- Yang 1967: Fermions, spin 1/2.

This piece becomes an algebraic problem, and the algebraic problem eventually resolves into two coupled Fredholm equations for two unknown functions, σ and ρ as a function of p:

$$2\pi\sigma = -\int_{-B}^{B} \frac{2c\sigma(\Lambda')d\Lambda'}{c^2 + (\Lambda - \Lambda')^2} + \int_{-Q}^{Q} \frac{4c\rho dp}{c^2 + 4(p - \Lambda)^2}, \tag{11}$$

$$2\pi\rho = 1 + \int_{-B}^{B} \frac{4c\sigma d\Lambda}{c^2 + 4(p - \Lambda)^2}. \tag{12}$$

If you solve these two equations, you will get all the properties of the ground state for the fermion problem without the harmonic trap.

Now we apply that to the case with the harmonic trap (Fig. 13). And in this region, however, we now have this potential due to the trap, and the kinetic energy part we use the solution due to Bethe ansatz. So, with that, you can solve the problem completely. After that is done, you can show that, in fact, this conjecture was correct (Fig. 14). The Thomas–Fermi method gives this whole curve, and we have now a solution. This is not a sketch, this is the exact calculation.[1–3] The calculation is, of course, based on solution of the Fredholm equations. Of course, we do know how to analytically solve that, but to any degree of accuracy that you want, we can calculate the curve to better accuracy than you have assigned.

All these have appeared in these publication: Refs. 1–3, and in additional publications to come. We are also generalizing it to finite temperature. The trick is because for finite temperature without delta function, that problem has also been

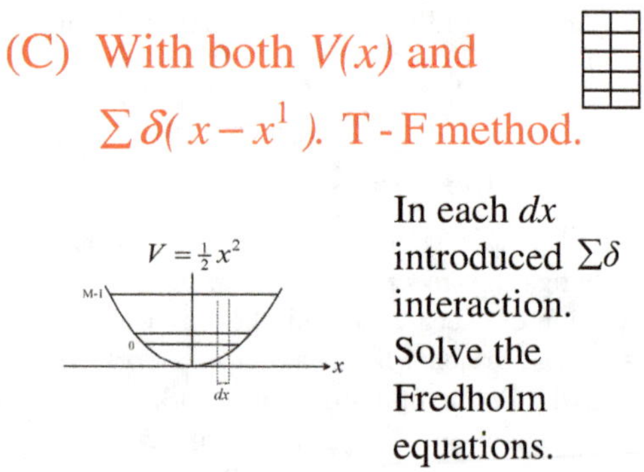

Fig. 13.

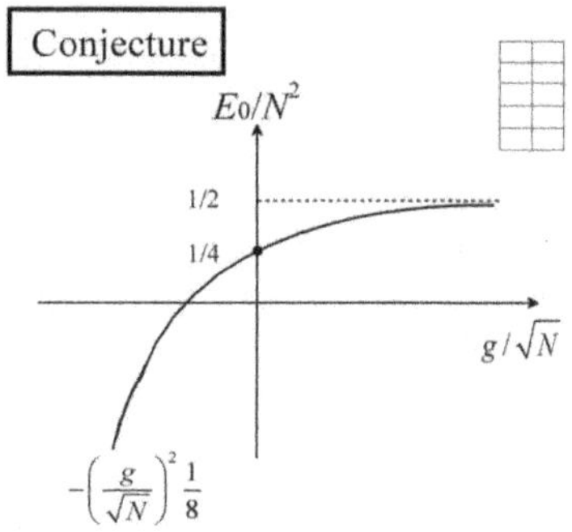

Fig. 14.

solved in the 1960s. And so, we just use that coupled with the Thomas–Fermi method, then we will get the finite temperature solution for the original Hamiltonian problem.

References

1. Z.-Q. Ma and C. N. Yang, *Chin. Phys. Lett.* **26**, 120505 (2009).
2. Z.-Q. Ma and C. N. Yang, *Chin. Phys. Lett.* **26**, 120506 (2009).
3. Z.-Q. Ma and C. N. Yang, *Chin. Phys. Lett.* **27**, 020506 (2010).

https://doi.org/10.1142/9789819826636_0021

TOPOLOGY AND GAUGE THEORY IN PHYSICS*

CHEN NING YANG
Tsinghua University, Beijing, China
Chinese University of Hong Kong, Hong Kong, China

Received 15 November 2012
Published 4 December 2012

I am deeply touched by the short message that Tonomura recorded for all of us today. It reminded me of my first visit to his laboratory in the early 1980s and the many conferences that I attended that he had organized in Japan. It also recalled for me the many discussions that he and I had, not only about the Aharonov–Bohm effect, but also about flux movement in superconductors, especially after the 1987 discovery of high-temperature superconductors. He and I had in these years many warm and fruitful meetings in Japan, in China, and in the United States. Last winter I was very happy to have received a photograph of him standing in a garden a few months after his operation. I thought he was on his way to full recovery. But that was not to be.

My title today is "Topology and Gauge Theory in Physics." In 1959, Aharonov and Bohm[1] proposed what is now known as the Aharonov–Bohm effect. If you want to appreciate the full meaning of this very important proposal, you have to look back into Maxwell's papers[2–5] of the 19th century. In his papers and his books, you will find that the vector potential **A** was *always* there: At the *very beginning* of his studies, he tried to express Faraday's amorphous concept of the *electrotonic state* in definite mathematical form.[6] He conceived of it as a vector, which later acquired a name: the vector potential **A**. So it was natural that it played an important role in all of his papers. It was after his death that Heaviside and Hertz eliminated the vector potential. Heaviside was a very enthusiastic and brilliant engineer. He was extremely happy that he had succeeded in eliminating the vector potential.[7] He

*A speech given at a conference in memory of Tonomura, Tokyo, May 2012.

Reproduced from International Journal of Modern Physics A, Vol. 27, No. 30, 1230035 (2012), DOI: S0217751X12300359

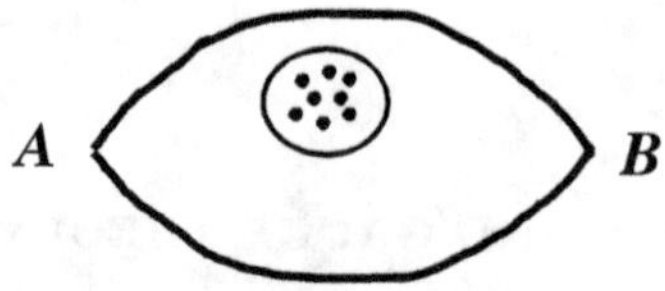

Fig. 1.

said for example,

> *The reader who is acquainted with (at present) the more "classical" method of treating the electromagnetic field in terms of the vector and the scalar potentials cannot fail to be impressed by the difference of procedure and of ideas involved. In the present method, we are, from first to last, in contact with those quantities which are believed to have physical significance, instead of with mathematical functions of an essentially indeterminate nature, and with the laws connecting them in simplest forms.*
>
> (Heaviside 1894)

Heaviside was not exactly wrong in saying that the vector potential was a mathematical function of an *essentially indeterminate* nature. But in today's language we understand it is indeterminate only in the traditional sense. In fiber bundle language the vector potential is a "connection." The subtlety of this concept resides *precisely* in its *transformable* character.

After Heaviside there was a *dogma* that the field strengths **E** and **H** are observable, and the vector potential **A** is not observable. It is only of mathematical interest and can be eliminated. This was written into all textbooks when I was a graduate student. It was drilled into our minds that the important things are **E** and **H**, the vector potential being superfluous. Then Aharonov–Bohm proposed an experiment shown in Fig. 1. An electron beam is split at point A and recombines at point B. A long solenoid at the center contains magnetic flux perpendicular to the plane of the paper.

If the flux is totally confined, then outside of the confinement, there is no **E**, no **H**. So the two paths of electron, starting from A to B, would not be subject to electromagnetic influences according to the traditional view. And so the image at B should not be dependent on the total flux inside of the solenoid. But Aharonov and Bohm in 1959 said the image at B *should be dependent* on the total flux.

The Aharonov–Bohm's proposal of 1959 was hotly debated. I remember I was in the Institute in Princeton at the time, and there were heated debates. One of the persons who absolutely refused to believe the Aharonov–Bohm experiment was Wigner. And those of us who knew Wigner knew it was absolutely impossible to change his opinion once he has formed a definite idea. That was, of course, part of his greatness as a physicist.

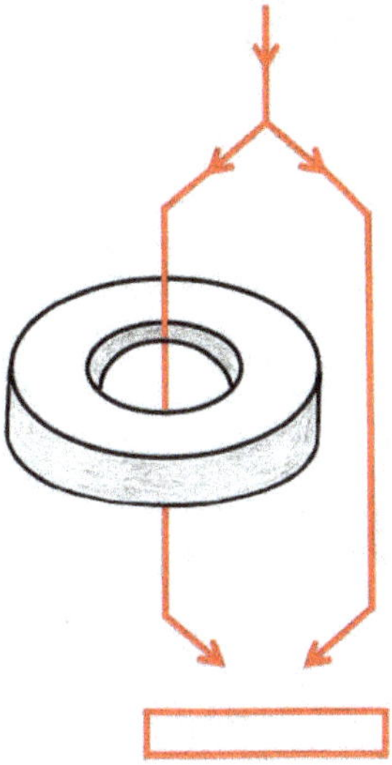

Fig. 2.

The A–B experiment was within a few years supposedly confirmed by Chambers.[8] But Chambers' experiment was not with a confined flux, but with a tapered needle. And there were endless discussions about what happened to the leaked flux around the tapered needle. It was finally the Tonomura experiments of 1983[9] and 1986[10] which settled the matter. With his beautiful electron holographic arrangements, he made the experiment which Dr. Zhu had just described to you.

This experiment of Tonomura and collaborators showed, first, the topological meaning of the vector potential in electromagnetism. Second, it showed that fluxes in superconducting rings are in quantized units of *ch* over 2*e*. Notice the 2. That 2 is, of course, because of the *pairing* of fermions in a superconductor, which was the essence of the existence of superconductivity. Thus Tonomura's experiment quantitatively and explicitly also showed, because of this fact of 2, the existence of Cooper pairs. This beautiful work is the first direct experimental proof of the importance of topology in physics.

I shall now go over a number of other developments, experimental and theoretical, in modern physics, which are deeply related to topological concepts, and to gauge transformations. The first was the Van Hove singularity.[11] In 1953, Van Hove was a "member," as I was, at the Institute for Advanced Study. At that time, there were lots of computations with the then available computers of the vibrational frequency distribution in crystals. Papers by several people at the University of Maryland showed that there were ups and downs in such distributions. Figure 3

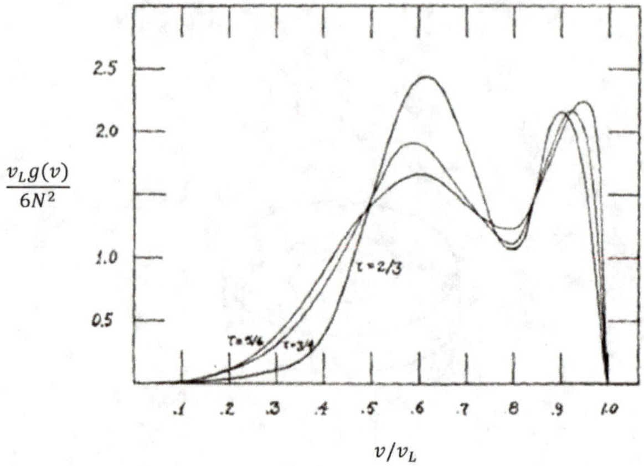

Fig. 3.

is reproduced from an early paper around that time. The abscissa is the frequency. The ordinate is the number of modes in each frequency range. (Such approximate computations were started, as you remember, already in 1905 and 1906 by Einstein, and later by Max Born. It was only in the 1950s, with computers, that people can make more accurate computations.) The resultant ups and downs caused quite a stir, as I vividly remember. It was Van Hove, who had more mathematical training than most of us, who pointed out that for topological reasons, because of Morse theory, there should in fact be singularities in the spectrum, not just ups and downs. The result of today's computations is shown below (Fig. 4).

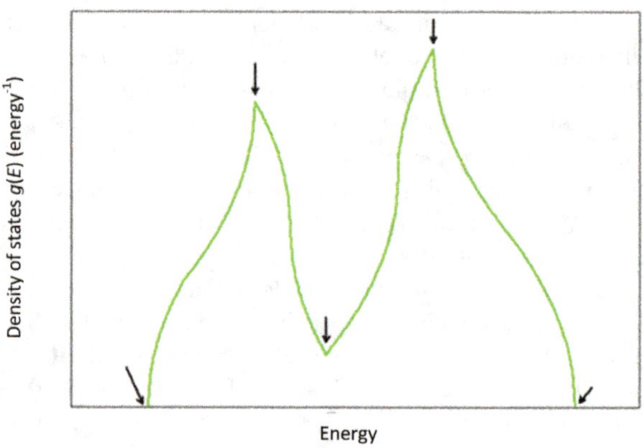

Fig. 4. A sketch of density of states $g(E)$ versus energy E for a simulated three-dimensional solid.

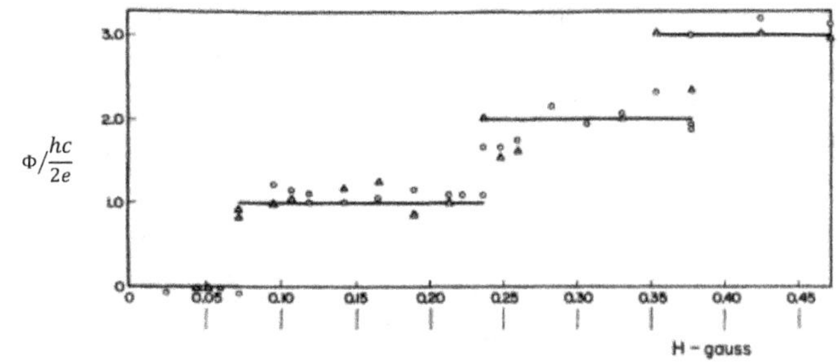

Fig. 5.

Morse theory is a generalization to higher dimensions of a simple topological theorem: A continuous function $F(x)$ of one periodic variable x has at least one maximum and one minimum. In higher dimensions there must also be saddle points. Morse theory played a very important role in the early development of topology in the 20th century, and Van Hove very elegantly pointed out that this is exactly what would explain those ups and downs in the frequency distribution in crystals.

Another important physical result related to gauge transformation and topology is the flux quantization experiment of 1961. This graph was from the original paper of Fairbank and his students[12] (Fig. 5).

The abscissa is the magnetic field. They had a superconducting ring and they were trying to measure the flux inside of it. This graph was in the first publication. If you look at today's publication, all those scattered points have coagulated into very good horizontal bars. The reason for this stepwise result is deeply related to gauge transformation and to topology.

Another very important idea which was finally understood in terms of topology, was Dirac's magnetic monopole of 1931.[13] I think everyone here knows this famous paper of 1931 of Dirac, who said that we have electrons with charge e; if there also exists a magnetic monopole with magnetic charge g, then

$$\frac{2eg}{\hbar c} = \text{integer}\,. \tag{1}$$

This was a paper which was very famous in 1931. It was also famous for several other reasons. For example, in the introduction of that paper, Dirac said the progress of physics, of theoretical physics, depends on two avenues of attack: one based on experiments, and the other based on beautiful mathematics. As time goes on, he said, the avenue based on experiments is going to become more and more difficult. So we have to search for advances in physics by looking into beautiful mathematics. It is extremely interesting today to read this 80 year old philosophy of Dirac's.

Dirac's quantization rule (1) turns out to be an elegant topological result. To describe how I realized this please allow me to first tell you a story.

In 1968–1969 I taught a course on general relativity at Stony Brook. One day at the lecture I copied on the blackboard the famous curvature equation of Riemann:

$$R^l_{ijk} = \frac{\partial}{\partial x^j}\begin{Bmatrix} l \\ ik \end{Bmatrix} - \frac{\partial}{\partial x^k}\begin{Bmatrix} l \\ ij \end{Bmatrix} + \begin{Bmatrix} m \\ ik \end{Bmatrix}\begin{Bmatrix} l \\ mj \end{Bmatrix} - \begin{Bmatrix} m \\ ij \end{Bmatrix}\begin{Bmatrix} l \\ mk \end{Bmatrix}. \tag{2}$$

As I did this it occurred to me that this equation looked very much like an equation in the 1954 papers of Yang–Mills:

$$F_{kj} = B_{k,j} - B_{j,k} + B_k B_j - B_j B_k\,. \tag{3}$$

After the lecture I compared the two equations, and quickly realized that in fact if I define matrices β_k and R_{jk} by

$$\langle i|\beta_k|l\rangle = \begin{Bmatrix} l \\ ik \end{Bmatrix}, \quad \langle i|R_{jk}|l\rangle = R^l_{ijk}\,, \tag{4}$$

then (2) is the $\langle i||l\rangle$ element of the matrix equation:

$$R_{jk} = \frac{\partial}{\partial x^j}\beta_k - \frac{\partial}{\partial x^k}\beta_j + \beta_k\beta_j - \beta_j\beta_k\,. \tag{5}$$

Equations (3) and (5) are thus *identical.* In excitement I went to consult Jim Simons who was then the chairman of mathematics at Stony Brook. He told me both equations are *curvature* equations in fiber bundle theory. Later at my request Simons gave us physicists a series of luncheon lectures about topology and fiber bundle theory. We thus understood *for the first time, that the mathematics of gauge theory, both Abelian and non-Abelian, had been independently developed by mathematicians in their beautiful fiber bundle theory.*

It is impossible to describe my *electrified* feeling upon realizing that electromagnetism, *a basic structure of the physical world*, is in fact based on subtle and beautiful mathematical constructs. In an article published later[14] I wrote:

> *In* 1975, *impressed with the fact that gauge fields are connections on fiber bundles I drove to the house of Shiing-Shen Chern in El Cerrito, near Berkeley. (I had taken courses with him in the early 1940's when he was a young professor and I an undergraduate student at the National Southwest Associated University in Kunming, China. That was before fiber bundles had become important in differential geometry and before Chern had made history with his contributions to the generalized Gauss–Bonnet theorem and the Chern classes.) We had much to talk about: friends, relatives, China. When our conservation turned to fiber bundles, I told him that I had finally learned from Jim Simons the beauty of fiber-bundle theory and the profound Chern–Weil theorem. I said I found it amazing that gauge fields are exactly connections on fiber bundles, which the mathematicians developed without reference to the physical world. I added "this is both thrilling and puzzling, since you mathematicians dreamed up these concepts out of nowhere." He immediately protested, "No, no. These concepts were not dreamed up. They were natural and real."*

Table 1. Gauge field terminology versus bundle terminology.

Gauge field terminology	Bundle terminology
Gauge (or global gauge)	Principal coordinate bundle
Gauge type	Principal fiber bundle
Gauge potential b^k_μ	Connection on a principal fiber bundle
S (Eq. (8))	Transition function
Phase factor Φ_{QP}	Parallel displacement
Field strength $f^k_{\mu\nu}$	Curvature
Source (electric) j^k_μ	?
Electromagnetism	Connection on a U_1 bundle
Isotopic spin gauge field	Connection on a SU_2 bundle
Dirac's monopole quantization	Classification of U_1 bundle according to first Chern class
Electromagnetism without monopole	Connection on a trivial U_1 bundle
Electromagnetism with monopole	Connection on a nontrivial U_1 bundle

Also in 1975 T. T. Wu and I published a *dictionary*[15] of the terms used in physicists' gauge theory and the corresponding ones used in mathematicians' fiber bundle theory.

One entry, source J^k_μ, is the charge-current 4-vector, which is, of course, of *fundamental importance* to physicists. But Simons told us "we do not consider such quantities," and that explains the "?" in the dictionary.

Why did the mathematicians not consider J^k_μ? I think there may be a subtle reason: In 1944 fiber bundle theory became important in topology with Chern's generalization of the Gauss–Bonnet theorem to four dimensions. Chern considered an integral of

$$f^k_{\mu\nu}\tilde{f}^k_{\mu\nu}\,, \tag{6}$$

where $\tilde{f}^k_{\mu\nu} = \varepsilon_{\mu\nu\alpha\beta}f^k_{\alpha\beta}$, is the dual of $f^k_{\mu\nu}$. (It exchanges **E** and **H**.) Now conceptually J_μ results from a variation of the Lagrangian which is the integral of

$$f^k_{\mu\nu}f^k_{\mu\nu}\,. \tag{7}$$

I think mathematicians were so fascinated with (6) that they neglected to consider expression (7) which is of *primary importance* to physicists.

Be that as it may, when I. M. Singer visited Stony Brook I showed him the Wu–Yang paper.[15] He brought it to Oxford. The result was a paper by Atiyah, Hinchin and Singer[16] which dealt with *sourceless* gauge fields, i.e. gauge fields for which

$$J^k_\mu = 0\,, \tag{8}$$

and for which the integral of expression (7) reaches an extremum. It is no exaggeration to say[17] that this three-man paper was one of the catalysts that ushered in the close collaboration between physicist and mathematicians in the last thirty odd years.

We now return to Dirac's quantization rule (1). Topologically it originated from the subtle structure of the vector potential around a magnetic monopole. When I showed it to Simons, he said "Dirac had anticipated Chern by more than a dozen years."

"Anomaly" was in a way first discovered in 1949 by the famous experimental physicist Steinberger when he calculated the rate of the process $\pi^0 \to \gamma + \gamma$. π^0 is a pseudoscalar and this process is related to expression (6) above. Much later it became clear that such anomalies are inherent in the topology of gauge fields.

More recently there are great excitements about topological insulators. One source of its remarkable properties is related again to expression (6) above. I think it is safe to predict that topology will continue to play new and subtle roles in physics.

Except for the Van Hove singularity, all of the examples mentioned above of the entry of topology into physics are basically related to the mathematical structure of gauge theory. It is *truely amazing*[a] that:

(1) Two hundred years ago, Michael Faraday (1791–1867), who knew little mathematics, was able to conceive of a vague and somewhat slippery concept of the *electrotonic state.*
(2) Maxwell (1831–1879) was able to formulate it into a mathematical quantity which was called the vector potential.
(3) The conceptual beauty of this quantity was independently discovered by the mathematicians *without reference to the physical world*, and was called the connection.

References

1. Y. Aharonov and D. Bohm, *Phys. Rev.* **115**, 485 (1959).
2. J. C. Maxwell, *Trans. Camb. Phil. Soc.* **10**, 27 (1856) [in *Scientific Papers of J. C. Maxwell*, Vol. 1, ed. W. D. Niven (Dover Publications, 1890), p. 155].
3. J. C. Maxwell, *Phil. Mag.* **21**, 281, 338 (1861).
4. J. C. Maxwell, *Phil. Mag.* **23**, 12, 85 (1862) [in *Scientific Papers of J. C. Maxwell*, ed. W. D. Niven (Dover Publications, 1890), p. 451].
5. J. C. Maxwell, *Phil. Trans. R. Soc. (London)* **155**, 459 (1865) [in *Scientific Papers of J. C. Maxwell*, ed. W. D. Niven (Dover Publications, 1890), p. 526].
6. A. C. T. Wu and C. N. Yang, *Int. J. Mod. Phys. A* **21**, 3235 (2006).
7. O. Heaviside, *Electromagnetic Theory* (The Electrician Printing and Publishing Company, 1893).
8. R. G. Chambers, *Phys. Rev. Lett.* **5**, 3 (1960).
9. A. Tonomura, T. Matsuda, R. Suzuki, A. Fukuhara, N. Osakabe, H. Umezaki, J. Endo, K. Shinagawa, Y. Sugita and H. Fujiwara, *Phys. Rev. Lett.* **48**, 1443 (1982).

[a]I had commented on these amazing historical facts in a talk in Tokyo.[18]

10. A. Tonomura, N. Osakabe, T. Matsuda, T. Kawasaki, J. Endo, S. Yano and H. Yamada, *Phys. Rev. Lett.* **56**, 792 (1986).
11. L. Van Hove, *Phys. Rev.* **89**, 1189 (1953).
12. B. S. Deaver, Jr. and W. M. Fairbank, *Phys. Rev. Lett.* **7**, 43 (1961).
13. P. A. M. Dirac, *Proc. R. Soc. London, Ser. A* **133**, 60 (1931).
14. C. N. Yang, *Phys. Today* **33**, 42 (1980).
15. T. T. Wu and C. N. Yang, *Phys. Rev. D* **12**, 3845 (1975).
16. M. F. Atiyah, N. J. Hinchin and I. M. Singer, *Proc. R. Soc. London, Ser. A* **362**, 425 (1978).
17. I. M. Singer, *Proc. Symp. Pure Math.* **48**, 198 (1988).
18. C. N. Yang, *JJAP Series* **9**, 3 (1993).

https://doi.org/10.1142/9789819826636_0022

Early examples of topological concepts in physics*

C. N. Yang
Institute for Advanced Study, Tsinghua University, P. R. China

Accepted 27 June 2018
Published 29 June 2018

In the mid-1940s, S. S. Chern published an "intrinsic proof" of a generalization of the Gauss–Bonnet Theorem to 4-dimensions. The paper led to the Chern Class and Chern Numbers, to the new exciting field of global differential geometry, and to new important topological concepts in other areas of mathematics. Andrei Weil was one mathematician who was greatly impressed. He wrote an enthusiastic review of the paper which became very influential.

A few years later, in 1946–1949, several totally unexpected new elementary particles were discovered by experimental physicists. They were of different kinds, with very different quantum numbers, and quickly became physicist's center of attention.

One day in 1948, I was present at a lunch in which Weil told Fermi his speculation that these new particles might be related to some topological classification ideas in geometry. Neither Fermi, nor I, nor others at that lunch, understood what Weil had meant that day by his speculation across the boundary of math–physics.

Many years later, in the mid-1970s, after I learned from Jim Simon elements of fiber bundle geometry and related concepts, I realized Weil maybe speculating that day about possible relationships between the new particles plus their new quantum number with topological concepts such as the Chern Numbers. For details please see Ref. 1.

****

In an article published in 2012,[2] I discussed in some detail the following early entry of topology into physics:

- The Aharonov–Bohm experiment proposed theoretically in 1959, and verified experimentally by Tonomura in 1983–1986.

*This article will also appear in "*Topology and Physics*", edited by C. N. Yang, M. L. Ge and Y. H. He (World Scientific, 2018).

Reproduced from Modern Physics Letters A, Vol. 33, No. 22, 1830009 (2018), DOI: 10.1142/S0217732318300094

- In the early 1950s physicists used the new computers to calculate the vibrational frequency distribution of crystals, and were surprised to find unexplained ups and downs in the spectra. Were they real? Or just quirks of the computation? The puzzle was resolved in a 1953 paper of Van Hove which introduced topology, viz. Morse Theory, into physics.

****

That topological concepts are important in physics is now well known, especially in phenomena/problems involving Abelian or non-Abelian phases. Here is an example which shows that in one problem in *classical* Maxwell theory, topology already plays an essential role:

Consider

An EM field interacting with both an electric charge e and a magnetic charge g,

a problem which had been considered by Dirac in 1931.[3] The electromagnetic potential (i.e. the connection), when analytically continued, forms a complicated ***nontrivial*** manifold. The action integral $\boldsymbol{a}$ is then definable ***only modulo*** $\boldsymbol{4\pi eg}$.[4]

If one tries to quantize this theory, ala Feynman's path integral, one would be dealing with the quantity

$$\exp(ia/\hbar)\,,$$

which is meaningful only if

$$2eg/\hbar = \text{an integer}\,.$$

This condition, first given by Dirac, is thus a *consequence* of the topology of classical Maxwell theory.

References

1. C. N. Yang, *Phys. Today* **65**, 33 (2012).
2. C. N. Yang, *Int. J. Mod. Phys. A* **27**, 1230035 (2012).
3. P. A. M. Dirac, *Proc. R. Soc. London A* **133**, 60 (1931).
4. T. T. Wu and C. N. Yang, *Phys. Rev. D* **14**, 437 (1976).

https://doi.org/10.1142/9789819826636_0023

C. N. Yang on Symmetry and Physics*

Published 28 February 2019

A poem by a famous poet in the Sung Dynasty shows symmetry: it could be read forward and backward. A piece from Bach ("the Crab Canon") shows temporal symmetry.

Symmetry has long been exploited for its beauty and science. The ancient Shang in China used symmetry in their bronze vessels. Spanish mosques exhibit lots of symmetry. In fact, patterns used in Islamic culture fully reflect the complete symmetry of reflections and rotations on a two-dimensional plane. There are only 17 such wall paper symmetry groups (see inset on tessellation). A poem by a famous poet in the Sung Dynasty shows symmetry: it could be read forward and backward. A piece from Bach ("the Crab Canon") shows temporal symmetry.

By unraveling the historical exploitation of symmetry from the Greeks to the modern age, Professor Yang showed how the concept of symmetry has significantly contributed to the concept of invariance; and how the study of global and more importantly local symmetry has enhanced our understanding of gauge field. Symmetry can be defined as the invariance in the pattern that is observed when some transformation is applied to it. Normally, we associate symmetry with geometrical forms, like the vase or the patterns in the mosques. Although the concept of symmetry had its origin in geometry, it is possible to extend invariance with respect to transformation of other kinds. An example is electromagnetic interactions in which the force remains unchanged if the positive and negative charges are interchanged.

*This article was written by Prof. L. C. Kwek based on the talk delivered by Prof. C. N. Yang at NTU, Singapore in July 2006. [An earlier version of this article appeared in the "Institute of Advanced Studies at NTU," issue 3, January 2007.]

Reproduced from International Journal of Modern Physics A, Vol. 34, No. 05, 1930004 (2019), DOI: 10.1142/S0217751X19300047

Prof. C. N. Yang

Indeed the concept of symmetry has been applied extensively in physics for a long time. Albert Einstein, for instance, formulated his special theory of relativity from the symmetry of space and time. The laws of electromagnetism on which the special theory of relativity rests are governed by Lorentz symmetry. In fact, Einstein resisted the idea of Lorentz symmetry but he subsequently embraced the idea and even generalized it into the theory of General Relativity.

In the realm of microscopic particles like atoms and molecules, certain discrete symmetries like the spatial reflection symmetry or parity (P), time reversal (T) and charge conjugation (C) were found to be extremely useful. Both parity and time reversal symmetry was introduced in quantum physics by Eugene Wigner, where important spectroscopic results were explained for the first time on the basis of a group-theoretic treatment of permutation, rotation and reflection symmetries. In 1931, symmetry due to charge conjugation was introduced by P. A. M. Dirac to account for the existences of antiparticles.

The laws governing gravity, electromagnetism, and the strong interaction are invariant with respect to C, P and T independently. However, in 1956 Lee Tsung-Dao and C. N. Yang pointed out that β-decay, a decay characterized by the emission of electrons and governed by the weak interaction, had not yet been tested for invariance under parity. Soon afterwards Chien-Shiung Wu and her colleagues performed an experiment showing that the weak interaction violates parity. Nevertheless, β-decay respects the combination of C and P as symmetry.

Parity and time reversal are discrete symmetries of nature. Space symmetry on the other hand is a continuous transformation. An important implication of symmetry is the existence of conservation laws. An important discovery that led eventually to the concept that "symmetry dictates interaction" was formulated by C. N. Yang and Robert Mills in the early seventies in the form of gauge theory.

Electromagnetism is an example of gauge theory. However, electromagnetism is an Abelian gauge theory. By extending the theory to a non-Abelian gauge theory, they were able to explain the symmetry governing protons and neutrons.

The secret of nature is symmetry but most observations in nature do not exhibit symmetry. A profound way of hiding symmetry is the phenomenon of spontaneous symmetry breaking. Thus, one needs a combination of gauge symmetry and symmetry breaking mechanism to explain results from experiment.

Three recent advances in symmetry were proposed during the period 1972 to 1984: in 1973, supersymmetry was mooted, in 1976, supergravity was formulated and in 1984 superstring was developed. These theories have attracted many bright students for the last twenty years. They are deeply entangled with modern mathematics and one of the pioneers in this field, Edward Witten has surprised many mathematicians many times using ideas developed initially in physics.

Tessellation of Space

Mathematicians have always been fascinated by patterns. Classifications of planar patterns and spatial lattices began seriously in the nineteenth century. A number of classification methods were developed and one of the ways to study patterns and lattices is to analyze the transformations that leave these patterns invariant through the concept of symmetry groups. For one dimension, there are seven such groups. For two dimensions, all planar tessellations can be classified completely by the 17 different plane symmetry groups.

Surprisingly, mathematicians discovered the complete classification of three-dimensional crystallographic or repeated patterns before they published the results for the easier two-dimensional problem. The complete list of 230 three-dimensional crystallographic patterns was completed in 1890 by Evgraf Stepanovich Fedorov and coworkers. It was only a year later that they worked out the details for the two-dimensional groups.

In 1944, Edith Muller began to study patterns in the Islamic art. It turns out that all 17 types of two-dimensional repeated patterns were used creatively in Islamic culture. In fact, in her pioneering work in 1944, she could only identified 11 of the 17 classes. It was not until 1987 that mathematicians were able to document all 17 types in the incredibly beautiful artistry achieved by the builders of the Alhambra Palace in Spain.

The Facade of Comares Palace (Alhambra, Spain)

When David Hilbert addressed the International Congress of Mathematicians in Paris in 1900, he proposed a number of outstanding problems that mathematicians should study. One of these problems was related to an understanding of higher dimensional crystallographic groups. In 1910, Ludwig Bieberbach proved that there are only finitely many such groups in any dimension.

In 1948, Hans Zassenhaus provided an algorithm to determine the complete set of representatives for space groups in arbitrary dimensions. By now, it is known that there are 4783 symmetry groups in four dimensions. However, the issue of enumerating the exact number of symmetry groups in all dimensions remains an outstanding and challenging problem.

Technical Research Papers

https://doi.org/10.1142/9789819826636_0024

SHOULD THERE BE KNO SCALING FOR e^+–e^- 2-JET EVENTS?†

T. T. CHOU
Department of Physics and Astronomy, University of Georgia, Athens, GA 30602, USA

CHEN NING YANG
Institute for Theoretical Physics, State University of New York, Stony Brook, NY 11794-3840, USA

Received 28 April 1986

Background reasoning is given for our recent prediction that KNO scaling does not obtain for e^+–e^- two-jet events.

1. Hadron-Hadron Collisions

1.1. *KNO scaling*

When KNO scaling[1] was proposed in 1972 we found it difficult to believe that it could be correct. The main reason for this attitude was the fact that with KNO scaling

$$\Delta n \propto \bar{n}, \tag{1}$$

which means that the distribution is broad. In contrast, for a stochastic process typically

$$\Delta n \propto \sqrt{\bar{n}}, \tag{2}$$

which means that the distribution is very narrow compared to KNO scaling. It was difficult for us to believe at that time that with many dozens of particles emitted the process could be nonstochastic like (1).

However, the experiments[2] of 1981 at the CERN collider showed that indeed KNO scaling was approximately correct. [The recent negative binomial fits to the multiplicity distribution mean that Eq. (1) is replaced by the statement that the ratio $\Delta n/\bar{n}$ increases with energy. In other words the recent negative binomial fits are *even broader* than a strict KNO scaling would require.]

Thus after 1981, experiments forced us to the view that some elements of the

† Article also to appear in *Second Asia-Pacific Conference*, Bangalore '86 (World Scientific, Singapore, 1986).
This work is supported in part by the US Department of Energy under grant No. DE-FG09-84ER 40160 and by the National Science Foundation under grant No. PHY 85-07627.

Reproduced from International Journal of Modern Physics A, Vol. 01, No. 02, pp. 415–420 (1986), DOI: 10.1142/S0217751X86000150

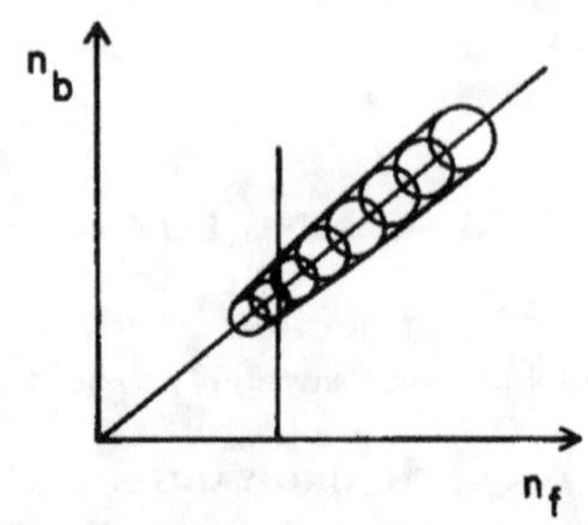

Fig. 1. Multiplicity distribution for hadron-hadron collisions at high energies. n_f and n_b refer respectively to the forward and backward charge multiplicities. We predict that for high energies the distribution becomes a long cigar as shown, which is identical to the prediction that $b \to 1$ [see Eq. (7)]. The cigar can be regarded as a collection of circles each corresponding to one angular momentum. The radius of a circle is $\sim\sqrt{n}$, because the width of the cigar is $\sim\sqrt{n}$. In other words, the fluctuations of n_f and n_b are $\sim\sqrt{n}$ for each angular momentum. We predict that[8] for each angular momentum the circle represents a product of two Poissons, one in n_f and the other in n_b.

dynamics of hadron-hadron collisions at high energies must be nonstochastic. Yet, it is difficult to believe that other elements of it could escape being stochastic. The question then was which aspect of the dynamics is stochastic and which nonstochastic?

1.2. *Separation of stochastic and nonstochastic aspects*

In 1983 the UA5 group published[3] a histogram of the $n_f - n_b$ (2-dimensional) distribution of events at the collider, where n_b and n_f, respectively, referred to backward and forward charge multiplicities. KNO scaling means that the distribution of $n = n_b + n_f$ satisfies KNO scaling. When one examines[3,4] the distribution (for fixed n) with respect to $n_f - n_b$ one finds that the distribution is a binomial, which, of course, is a stochastic distribution. Thus we reached the conclusion[4] that in the $n_f - n_b$ plane the distribution is stochastic in the northwest-southeast direction while it is nonstochastic in the northeast-southwest direction. We conjectured that this trend would continue to much higher energies. Since a stochastic distribution becomes very narrow relative to a nonstochastic distribution when $\bar{n} \to \infty$, we predicted that at every high energy the distribution in the $n_f - n_b$ plane would become a longer and longer cigar-like distribution (Fig. 1). That is, we predicted[4] that the parameter b would approach unity.

When that obtains, let us consider a fixed n_f as indicated by the vertical line in Fig. 1. The distribution of n_b along that line would be concentrated near a value of n_b which is close to n_f with a fluctuation of the order of $\sqrt{n_f}$. In other words, along that vertical line in Fig. 1, the charge multiplicity distribution in the backward hemisphere would be stochastic.

1.3. *Energy partition and T_p*

Accepting the above conclusion, E. Yen and we raised[5] the following question: How about the energy and momentum distribution of the outgoing particles in the backward direction once we have fixed n_f so that we are on the vertical line in Fig. 1? The charge multiplicity in the backward direction is, according to what we just said, stochastic. It would then seem most natural that other distributions in the backward hemisphere should also be stochastic. Of course stochastic means stochastic subject to certain conditions. What are these conditions?

The following obviously are to be satisfied by the outgoing particles in the backward hemisphere:

(i) Energy conservation taking into account the energy taken away by the leading particles. The leading particle concept was first formulated[6] in cosmic ray physics a long time ago. While the total energy in the backward direction is E_0 (= 270 GeV at CERN), only a fraction h of this would be carried by the central region particles in the backward hemisphere. The remaining fraction $1 - h$ would be taken away by the leading particles in the backward direction.

(ii) Each emitted particle would carry a bremsstrahlung factor d^3p/E, which was first used in multi-hadron production[7] in the 1940s.

(iii) There is a transverse momentum-cutoff factor for each particle which we take from experimental results to be

$$g(p_\perp) = \exp(-\alpha p_\perp), \qquad \alpha = 5.25\ (\mathrm{GeV/c})^{-1}. \tag{3}$$

Subject to these conditions the simplest stochastic distribution of the particles in the backward hemisphere gives a probability of the form

$$\delta\left(\sum_i E_i - E_0 h\right)\prod_i (d^3p_i/E_i)g(p_{\perp i}), \tag{4}$$

where the summation and multiplication extend over all the particles, charged or neutral, in the central region (i.e., without leading particles) in the backward hemisphere. To evaluate the single-particle spectrum from Eq. (4) we use[5] the equivalence relationship between a microcanonical ensemble and a canonical ensemble. Equation (4) describes a microcanonical ensemble at a fixed energy. For large multiplicities this gives a single-particle canonical particle distribution

$$(d^3p/E)g(p_\perp)\exp(-E/T_p), \tag{5}$$

where T_p is a temperature-like parameter which we have called partition temperature.

Equation (5) gives the single-particle distribution along the vertical line in Fig. 1, (i.e., for a fixed n_f). Fixing n_f is approximately equivalent to fixing $n = n_f + n_b$, as Fig. 1 indicates. Thus we arrive at the conclusion that for each n there is a partition temperature T_p, which we determine[5] from the experimental angular distribution at that multiplicity. The good fit to experimental data indicates the usefulness of the idea of the partition temperature.

2. e^+–e^- Collisions

2.1. *Should there be KNO scaling for e^+–e^- collisions?*

The physical picture that emerged in the above for hadron-hadron collisions can be summarized as follows (Fig. 1): Hadron-hadron collisions at very high energies is a superposition of many collisions each represented by one of the circles. Each circle represents a collision at one impact parameter and therefore one angular momentum. For each angular momentum the multiplicity fluctuates very little[a] (i.e., only stochastically). The wide fluctuation of multiplicities characteristic of KNO scaling is due to the superposition of the many circles in Fig. 1. In other words, *KNO scaling in hadron-hadron collisions is due to the wide fluctuations of the total angular momentum.*

Indeed the total angular momentum for the 270 GeV on 270 GeV collision at the CERN collider ranges from 0 to $2000\hbar$, and this wide fluctuation of the angular momentum *must* produce a wide fluctuation of the multiplicity since collisions at angular momenta 0 and $2000\hbar$ are not expected to yield the same average multiplicities.

The angular momentum in e^+–e^- annihilation is, on the other hand, 0 or $\hbar$. There is little fluctuation and the $n_f - n_b$ distribution should be represented by only *one* circle. *We thus predict that for e^+–e^- annihilation KNO scaling does not obtain.*

Is this prediction[8] in agreement with experiments? Because the energy range of existing data is too small, we cannot give a conclusive answer to this question. However existing data, when viewed from the right angle, do seem to exhibit[8] characteristics that suggest that our prediction is correct.

(i) In Fig. 2 we reproduce for comparison multiplicity fluctuation plots for e^+–e^- and hadron-hadron experiments. We emphasize that for the former (left column) the peak is more left-right symmetrical than for the latter (right column). Also the left-column figures exhibit on the low multiplicity side of the peaks a region where the curve has positive second derivatives. Such regions are absent in the figures of the right column. These properties all mark the left column figures as Poisson-like, in contrast to the figures in the right column.

(ii) A left-right correlation parameter b has been measured, yielding

$$\begin{aligned} &b \approx 0, && (e^+\text{–}e^- \text{ annihilation}),^{9} \\ &b = 0.14\ (62\ \text{GeV}),\ 0.55\ (900\ \text{GeV}), && (p\bar{p}).^{10} \end{aligned} \tag{6}$$

According to our view for e^+–e^- annihilation, the $n_f - n_b$ distribution is a circle, i.e., a product of two Poisson distributions with respect to n_f and n_b. For such a product distribution one naturally expects $b = 0$. On the other hand, for hadron-hadron collisions the increasing value of b through the relationship[3]

$$\frac{1-b}{1+b} = \frac{\langle (n_f - n_b)^2 \rangle}{\langle (n_f + n_b)^2 \rangle} \tag{7}$$

[a] See caption of Fig. 1 for details.

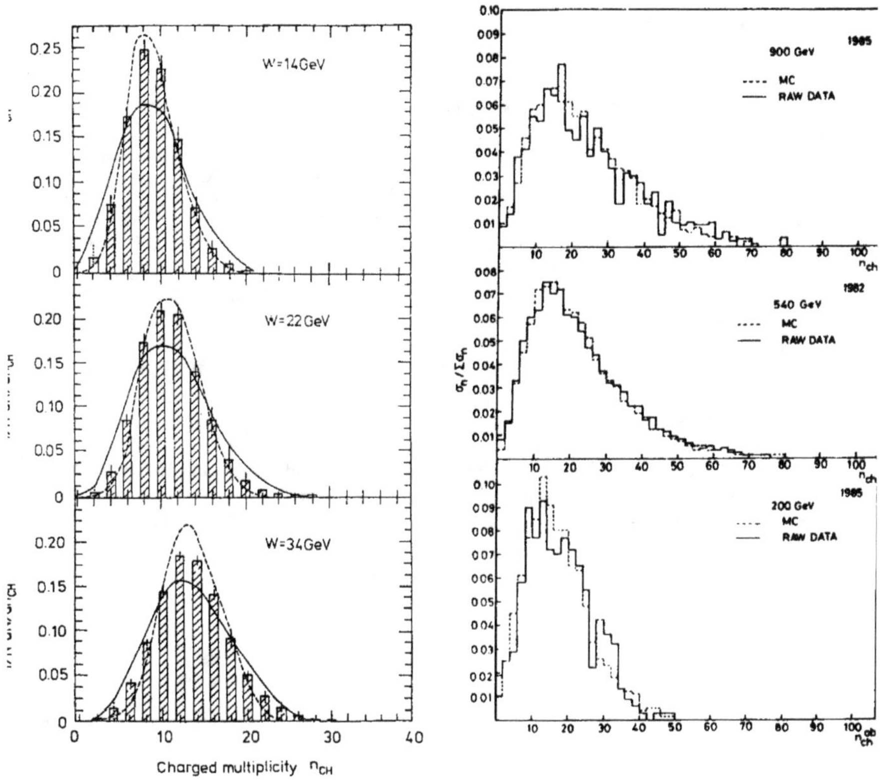

Fig. 2. Comparison of charged-particle multiplicity distribution for e^+–e^- annihilation and $p\bar{p}$ collisions. The e^+–e^- curves (left column) are reproduced from a TASSO publication.[11] No corrections for gluon bremsstrahlung effects have been made. The $p\bar{p}$ curves (right column) are reproduced from a UA5 report.[10] For details see these original papers.

implies that the distribution in the $n_f - n_b$ plane shrinks[4] towards a long cigar, as shown in Fig. 1.

2.2. *Partition temperature T_p in $e^+ - e^-$ annihilation*

Exactly as in hadron-hadron collision the same argument leads to the concept of a partition temperature for $e^+ - e^-$ annihilation. The only differences are: we consider only 2-jet events and ignore those with gluon emission; the transverse momentum now refers to the direction perpendicular to the jet axis. As there is only one angular momentum, there is now only one partition temperature T_p. We found that for $W = 14$ and 34 GeV, excellent fits[8] to the charged particle spectra are obtained with T_p equal to 1.6 and 3.3 GeV.

In recent publications[8] we have summarized our predictions for future hadron-hadron and $e^+ - e^-$ experiments.

References

1. Z. Koba, H. B. Nielsen and P. Olesen, *Nucl. Phys.* **B40** (1972) 317.
2. UA5 collaboration, K. Alpgard *et al.*, *Phys. Lett* **107B** (1981) 310; UA1 collaboration, G. Arnison, *et al.*, *Phys. Lett.* **107B** (1981) 320.
3. UA5 collaboration, G. Ekspong, Rome Conference (January 1983); UA5 collaboration, K. Alpgard, *et al.*, *Phys. Lett* **123B** (1983) 361.
4. T. T. Chou and Chen Ning Yang, *Phys. Lett* **135B** (1984) 175.
5. T. T. Chou, Chen Ning Yang and E. Yen, *Phys. Rev. Lett.* **54** (1985) 510; T. T. Chou and Chen Ning Yang, *Phys. Rev.* **D32** (1985) 1692.
6. G. Cocconi, *Phys. Rev.* **111** (1958) 1699.
7. H. W. Lewis, J. R. Oppenheimer and S. A. Wouthuysen, *Phys. Rev.* **73** (1948) 127.
8. T. T. Chou and Chen Ning Yang, *Phys. Rev. Lett.* **55** (1985) 1359; *Phys. Lett.* **167B** (1986) 453.
9. TASSO collaboration, DESY preprint 85-077 (August 1985).
10. UA5 collaboration, J. G. Rushbrooke, report at Multiparticle Conference Israel, June 1985, CERN EP/85-124.
11. TASSO collaboration, M. Althoff, *et al.*, *Z. Phys.* **C22** (1984) 307.

https://doi.org/10.1142/9789819826636_0025

A UNIFIED PHYSICAL PICTURE OF MULTIPARTICLE EMISSION: WIDE MULTIPLICITY DISTRIBUTION FOR $\bar{p}p$ AND NARROW MULTIPLICITY DISTRIBUTION FOR e^+e^- COLLISIONS*

T. T. CHOU
Department of Physics, University of Georgia, Athens, Georgia 30602, USA
and
CHEN NING YANG
Institute for Theoretical Physics, State University of New York, Stony Brook, New York 11794, USA

Received 2 July 1987

A unified physical picture for multiparticle emission in hadron-hadron and e^+e^- collisions is presented. It is based on the wide range of values of the total angular momentum in hadron-hadron collisions. Experimental support is reviewed. An extension of the considerations to give a theory of the momentum distribution of the outgoing particles is compared with experiments.

Introduction

In a high energy $\bar{p}p$ collision in the 540-GeV $S\bar{p}pS$ Collider at CERN, the total angular momentum of the collision ranges from 0 to $\sim 2000\hbar$ ($\cong 1.5\,\text{fm} \times 270$ GeV). This wide range of total angular momenta produces in elastic scattering a coherent superposition which manifests itself in the very high forward elastic peak observed in all high energy experiments which have been so far performed. A theory of elastic scattering based on such a concept was advanced in 1967–1968,[1] which was found to be in very good general agreement with experiments.[2] Additional consequences of this theory for diffraction dissociation, limiting fragmentation and other phenomena have been developed in the last twenty years.[3]

In the present paper we summarize our work in the last few years which focuses on the idea that the wide range of angular momenta in high energy hadron-hadron collision is the *cause* of the observed broad multiplicity distribution which has been fitted with KNO-scaling or negative binomials. A natural prediction of our view is that in e^+e^- collisions at high energy where the total angular momentum hardly fluctuates (being 0 or $1\hbar$), the multiplicity distribution is narrow and Poisson-like. If this prediction is right, then one has a *unified physical picture* of multiplicity distribution in hadron-hadron and e^+e^- collisions. We shall present experimental results which strongly support this prediction.

* Also to be published in review volume of *Multiparticle Production,—Directions in High Energy Physics*, ed. P. Carruthers (World Scientific, Singapore).

Reproduced from International Journal of Modern Physics A, Vol. 02, No. 06, pp. 1727–1753 (1987), DOI: 10.1142/S0217751X87000892

A natural extension of the physical picture mentioned above can be applied to the momentum distribution of emitted particles in the central region[4] (i.e., nonleading particles). Out of such considerations the concept of "partition temperature" T_p was developed. This concept will be discussed together with its application to the angular and momentum distributions of central region particles in both $\bar{p}p$ and e^+e^- collisions.

1. Multiparticle Production in Hadron-Hadron Collisions

1.1. *Multiplicity distribution*

A salient feature for multiparticle particle production processes in the central region in $\bar{p}p$ collisions observed in the early CERN $S\bar{p}pS$ Collider experiments at 540 GeV is the large fluctuation[5] in multiplicity distribution P_n, which was shown to satisfy approximately the (KNO)[6] scaling law

$$\bar{n}P_n = \psi(Z) \tag{1.1}$$

where Z = the KNO-variable $\equiv n/\bar{n}$. More recent measurements and analyses[7] indicate that the multiplicity distributions seem to fit better the negative binomial (NB) form

$$P_n = \frac{\Gamma(n+k)}{\Gamma(n+1)\Gamma(k)}\left[\frac{\bar{n}}{\bar{n}+k}\right]^n\left[\frac{k}{\bar{n}+k}\right]^k \tag{1.2}$$

where the adjustable parameter k (= 4.6 to 3.2 in the CERN Collider energy range) decreases with $\sqrt{s}$. Such distributions, KNO or NB, imply very large mean value of the fluctuation, $(\Delta n)^2 \equiv \overline{n^2} - \bar{n}^2$, for the observed charged-particle multiplicity: For KNO,

$$(\Delta n/\bar{n})^2 = \text{independent of energy}, \tag{1.3}$$

and for the negative binomial,

$$(\Delta n/\bar{n})^2 = (\bar{n})^{-1} + k^{-1} = \text{increasing with energy}. \tag{1.4}$$

We call these broad distributions *nonstochastic*. In contrast, a *stochastic* distribution, like a Poisson or binomial, gives typically

$$(\Delta n/\bar{n})^2 = 1/\bar{n} \tag{1.5}$$

which is narrow. Furthermore, a stochastic distribution will become infinitely narrow compared to a nonstochastic one as $\bar{n} \to \infty$.

When KNO scaling was first proposed, we were reluctant to accept it, because for a high energy collision with the emission of dozens of particles, it was difficult to believe that the process could be as nonstochastic as the KNO distribution. After experiments

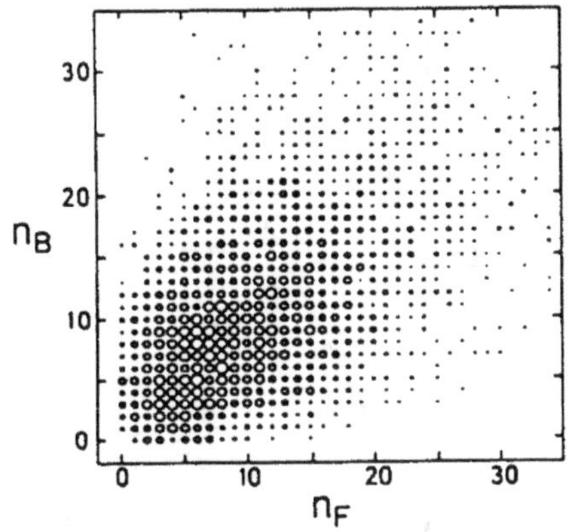

Fig. 1. Scatter plot of backward and forward multiplicities n_B and n_F for the η intervals $4 > \eta_F > 0$ and $0 > \eta_B > -4$ at 540 GeV. For each value of (n_B, n_F) the area of the circle is proportional to the number of events. (Reproduced from Ref. 8.)

at the CERN Collider confirmed in 1981 the broad approximate KNO (i.e., non-stochastic) distribution for the multiplicity, we were forced to *the view that some aspect of the dynamics of hadron-hadron collision is nonstochastic, while other aspects must be stochastic.* The question was then which aspects are stochastic and which are not.

1.2. *The stochastic and nonstochastic aspects of particle production processes*

1.2.1. *Separation of stochastic from nonstochastic aspects of collision*

The UA5 Collaboration published[8] in 1983 a probability distribution $P(n_F, n_B)$ of events for 540-GeV $\bar{p}p$ collisions with n_F and n_B denoting the forward and backward charge multiplicities respectively. Their scatter plot is reproduced here in Fig. 1. The exhibited distribution of events is analyzed[8] most conveniently in terms of the variables $n(= n_F + n_B)$ and $z(= n_F - n_B)$. Note that because of symmetry in pp and $\bar{p}p$ collisions, the probability distribution $P(n, z)$ must be even in the variable z.

If one sums the scatter plot over all values of z for a fixed n, and plots the result against n, one gets the experimental multiplicity distribution which is broad and nonstochastic. But if one keeps n constant and plots the probability of events against z, one gets[8,9] a binomial distribution which is, of course, a typical stochastic distribution. To see this, let us evaluate from the data exhibited in Fig. 1 the average value $\overline{z^2}$ for each n. The results fit very well the simple formula

$$[\overline{z^2} \text{ at a fixed } n] = 2n \tag{1.6}$$

which is the straight line shown in Fig. 2 against the experimental data points. This very good fit, especially the simple coefficient 2, suggests that the distribution with respect to z, at a fixed n, is a simple one. It turns out that in fact the simple binomial

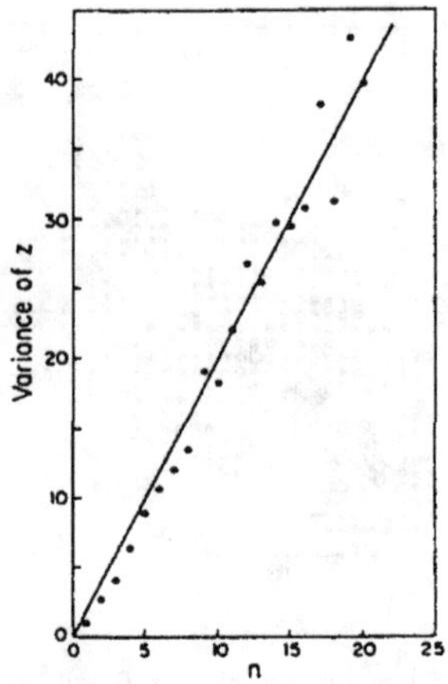

Fig. 2. Plot of the average value of z^2 for fixed n versus n. The straight line represents Eq. (1.6). (Reproduced from Ref. 9.)

$$P(n,z) = (\text{function of } n)C^{n/2}_{n_F/2}$$
$$= (\text{function of } n)C^{n/2}_{(n+z)/4}, \tag{1.7}$$

which does satisfy (1.6), gives an excellent fit as exhibited in Fig. 3.

Through the above analysis, we reached the conclusion that in the n_F–n_B plane, the $z(=n_F-n_B)$ distribution is binomial for each fixed n, which is of course stochastic, but the $n(=n_F+n_B)$ distribution is NB or approximately KNO, hence, nonstochastic. Combining (1.1) and (1.7), we have[9]

$$P(n,z) = \{(\bar{n})^{-1}\psi(n/\bar{n})\}\{[B(n)]^{-1}C^{n/2}_{(n+z)/4}\} \tag{1.8}$$

where $\psi(n/\bar{n})$ is the KNO scaling function $[(\bar{n})^{-1}\psi(n/\bar{n})$ may be replaced with the NB-distribution$]$ and $\bar{n}$ is the average total charged multiplicity. The normalization factor $[B(n)]^{-1}$ is inserted so that the expression $C^{n/2}_{(n+z)/4}$ summed over z in steps of $\Delta z = 2$ gives $B(n)$. The values of $B(2)$, $B(4)$, $B(6)$, $B(8)$, …, are $2+(4/\pi)$, $4+(32/3\pi)$, $8+(352/15\pi)$, $16+(1024/21\pi)$, … . Asymptotically, $B(n) = 2^{1+(n/2)}$ for large n. The clear separation exhibited in (1.8) of the stochastic from the nonstochastic aspects of multiparticle production processes gives conceptually a clear and attractive physical picture for high energy hadron-hadron collisions.

1.2.2. *Interpretation for the binomial distribution factor in $P(n,z)$*

The binomial form (1.7) for the multiplicity distribution along a constant $n(=n_F+n_B)$ line in the charge asymmetry variable $z(=n_F-n_B)$ may be realized in many different physical models. But the most intriguing aspect of (1.8) is the factor 2 which

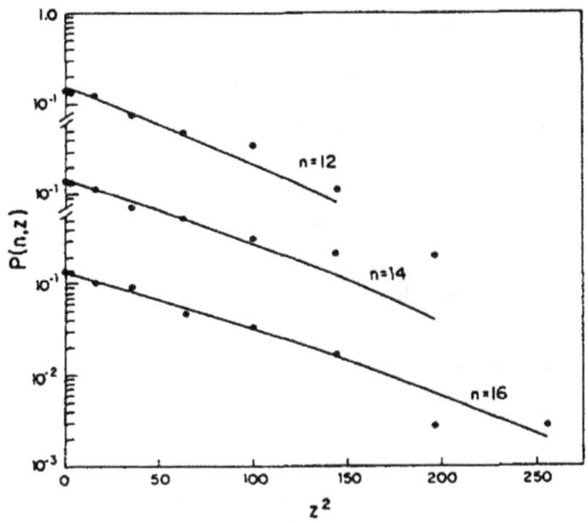

Fig. 3. Plot of the calculated z-distribution with Eq. (1.7) and the experimental data points in log scale for $n = 12, 14$, and 16. Both the theoretical curves and the experimental data points are normalized. (Reproduced from Ref. 9.)

came from (1.6). The simplest and most appealing physical picture that will explain this factor is one in which positive and negative charged particles are always produced together in pairs in either the forward or backward hemispheres. The positive charges, $n/2$ in number, are distributed in the forward and backward hemispheres in a binomial distribution $C_m^{n/2}$, where m is the number of positive charges on one side. The total charge is then assumed to be neutral on each side, so that $n_F = 2m$, and $n + z = 2n_F$. Hence, $m = (n + z)/4$, which yields the z-distribution in (1.7).

If the physical picture described above is approximately correct, then the net charge on either side, forward or backward, must be nearly neutral. This consideration leads us[10] to the prediction

$$\text{(root-mean-square net charge in each hemisphere)}/\sqrt{\bar{n}} \to 0, \text{ as } \bar{n} \to \infty \qquad (1.9)$$

which can be experimentally tested.

In the same spirit that one assumes the total charge to be small in each side, one would also assume the total isospin to be small in each side. The experimental consequence of this latter assumption is that in each hemisphere there will be approximately equal numbers of positive, negative and neutral particles.

1.2.3. *Cluster models*

The discussion above is one of the general category of cluster models[11] in which clusters are assumed to be emitted in the forward and backward hemispheres, each of which then splitting into k particles within the same hemisphere (or mostly within the same hemisphere). The model of Sec. 1.2.2 corresponds to the case $k = 2$. Extensive

Monte Carlo calculation[11] has been made by the UA5 group to fit $S\bar{p}pS$ data, including a discussion of possible range of values of k. We shall return to the value of k in Sec. 2 of this paper when we discuss e^+e^- collisions and compare it with $\bar{p}p$ collisions.

1.3. *Extrapolation of forward-backward multiplicity distribution to very high energies*

1.3.1. *Forward-backward multiplicity correlation*

The analysis in Secs. 1.1 and 1.2 of the probability distribution of events in 540-GeV $\bar{p}p$ collisions led us to the conclusion expressed in (1.8) that, in the n_F–n_B plane, the distribution is nonstochastic (KNO or NB) along the $n_F + n_B$ direction, and the distribution is stochastic (binomial) along the $n_F - n_B$ direction. At 540 GeV the n_F–n_B distribution is in the form of approximately a fat ellipse as shown in Fig. 1.

We have also emphasized that the widths of the two distributions behave very differently as $\bar{n}$ increases. [Compare Eqs. (1.3) and (1.4) with (1.5)]. Thus, if (1.8) should remain valid at higher and higher energies, the population of events would change drastically. Because a stochastic distribution becomes increasingly narrower relative to a nonstochastic distribution when $\bar{n}$ increases, the population in the n_F–n_B plane would become more and more concentrated in an elongated region along the $n_F = n_B$ axis. In other words, the contour lines for $P(n_F, n_B)$ will evolve from (approximately) fat ellipses at 540 GeV to much thinner ellipses as energy increases.

To analyze this problem quantitatively, we introduce a parameter α which measures the elongatedness of the two dimensional n_F–n_B distribution

$$\alpha \equiv \frac{\text{population average of } (n_F + n_B)^2}{\text{population average of } (n_F - n_B)^2} \tag{1.10}$$

or

$$\alpha \equiv \frac{\text{variance of } n}{\text{variance of } z}. \tag{1.11}$$

$\sqrt{\alpha}$ has a simple geometrical meaning. If the contour lines are exactly ellipses, $\sqrt{\alpha}$ is the ratio of the major-axis to the minor-axis. Now, using the relation

$$[\text{variance of } z] = 2\bar{n} \tag{1.12}$$

which is obtained by averaging both sides of (1.6) over all values of n, we get from (1.11),

$$\begin{aligned} \alpha &= [\text{variance of } n]/(2\bar{n}) \\ &= (\Delta n)^2/(2\bar{n}) \\ &\to \infty \qquad \text{as } \bar{n} \to \infty\,. \end{aligned} \tag{1.13}$$

For the last step, use is made of (1.3) and (1.4).

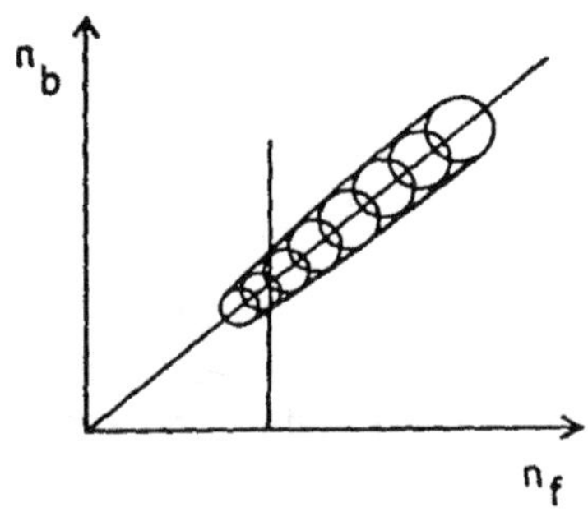

Fig. 4. Multiplicity distribution for hadron-hadron collisions at high energies. n_f and n_b refer respectively to the forward and backward charge multiplicities. We predict that for high energies the distribution becomes a band as shown, which is identical to the prediction that $b_c \to 1$ (see Eq. (1.17)). The band can be regarded as a collection of circles each corresponding to one angular momentum. The radius of a circle is $\sim\sqrt{n}$, because the width of the band is $\sim\sqrt{n}$. In other words, the fluctuations of n_f and n_b are $\sim\sqrt{n}$ for each angular momentum. We predict that for each angular momentum the circle represents a product of two Poissons, one in n_f and the other in n_b.

Equation (1.13) states that, *at very high energies, the population of events will concentrate in an extremely narrow band.*

When that is obtained, let us consider a fixed n_F as indicated by the vertical line in Fig. 4. The distribution of n_B along that vertical line would be concentrated near a value of n_B which is close to n_F with a fluctuation of the order of the width of the narrow band, viz., $\sim\sqrt{n}$ or $\sim\sqrt{n_F}$. Explicitly,

$$n_B = n_F + O(n_F^{1/2}). \tag{1.14}$$

Thus, in a high energy collision the forward and backward multiplicities n_F and n_B will exhibit *strong correlation.* Take a case when $\bar{n} = 200$, say. Then $\bar{n}_F = \bar{n}_B = 100$. n_F may fluctuate widely, say, from 50 to 200. But for each n_F, the backward multiplicity,

$$n_B \cong n_F \pm \sqrt{n_F}$$

fluctuates very little.

A parameter b_c is commonly used to characterize the correlation of multiplicities in the forward and backward hemispheres, and is defined by

$$b_c \equiv \frac{d\bar{n}_B}{dn_F}$$

$$= \frac{\overline{n_F n_B} - \bar{n}_F \bar{n}_B}{\overline{n_F^2} - \bar{n}_F^2}. \tag{1.15}$$

(In the literature, the symbol b is usually used. We use b_c here to distinguish it from the impact parameter b.) It is obvious that parameters b_c and α are related

$$\alpha = \frac{1 + b_c}{1 - b_c}. \tag{1.16}$$

Thus, our prediction (1.13) that the population will become more and more concentrated in a narrow band as $\bar{n}$ increases means that the correlation parameter b_c for hadron-hadron collisions will increase with energy and approaches unity in the ultrahigh-energy limit

$$b_c \to 1 \qquad \text{as } \bar{n} \to \infty. \tag{1.17}$$

Is this prediction in agreement with experimental data? We list below some measured b_c values[12] for pp and $\bar{p}p$ collisions over the ISR and $S\bar{p}pS$ energy range

$$\begin{aligned} b_c = 0.14 &\quad \text{at } \sqrt{s} = \ 62\ \text{GeV} \\ 0.44 &\qquad\qquad 200\ \text{GeV} \\ 0.57 &\qquad\qquad 546\ \text{GeV} \\ 0.63 &\qquad\qquad 900\ \text{GeV}. \end{aligned} \tag{1.18}$$

The trend is clearly consistent with our conjecture.

1.3.2. *Incoherent superposition of collisions at different impact parameters each with narrow multiplicity fluctuations*

Accepting the above conclusion that at ultrahigh energies, both n_F and n_B fluctuate widely, but in a correlated way, with $n_F \cong n_B$, what can be the underlying physical cause for such phenomena?

A natural possibility for such an underlying cause is the wide fluctuation of the total angular momentum, or the impact parameter b, already mentioned in the Introduction above. It is reasonable to assume that for a small impact parameter ($b \cong 0$), i.e., a head-on collision, the multiplicity is large, while for a large impact parameter, say $b > 1.5$ fm, the multiplicity is small,[13] i.e., the multiplicity is assumed to depend on b. Since both sides, forward and backward, *share* the same impact parameter b, this assumption leads naturally to $n_F \cong n_B$. Proceeding along such reasoning, one arrives at the picture illustrated in Fig. 4 in which for each fixed b, the multiplicity fluctuations of both n_F and n_B are stochastic, represented by a circle in Fig. 4. The superposition of such stochastic distributions gives the long narrow band which exhibits wide fluctuations in n ($= n_F + n_B$).

In short, we propose a physical mechanism for approximate KNO scaling in $\bar{p}p$ collisions: *The experimentally observed broad multiplicity distribution (KNO or NB) is the result of an incoherent superposition of collisions at different impact parameters, each of which gives a narrow multiplicity distribution.*

What should be the precise form of the narrow multiplicity distribution for fixed impact parameter which is represented by a circle in Fig. 4? If one assumes this distribution to be a product of two Poissons, one in n_F and one in n_B, both with average $n_0(b)/2$

$$f(n_0/4, n_F/2)f(n_0/4, n_B/2), \tag{1.19}$$

where

$$f(a, m) = \exp(-a)a^m/m!, \tag{1.20}$$

then one has a very simple explanation of the binomial factor in (1.7) because

$$f(n_0/4, n_F/2)f(n_0/4, n_B/2) = 2^{-n/2}f(n_0/2, n/2)C^{n/2}_{n_F/2}. \tag{1.21}$$

The superposition of distributions (1.21) for different b's, each with its own value for $n_0 = n_0(b)$, gives the n-dependent factor of (1.7). Explorations along this direction is in progress. We remark here that the concept of an average multiplicity n_0 which depends on the impact parameter b has been discussed in the literature before.[13] The assumption that for a fixed b the multiplicity distribution is a product of two Poissons (1.19) is recent.[14]

1.4. *Energy partition, partition temperature and single-particle momentum distribution*

1.4.1. *Stochastic energy partition and single-particle spectrum*

If eventually the population of events clusters into a very narrow band at very high energies and very large average multiplicities $\bar{n}$, the distribution with respect to n_B, for a fixed n_F, would be concentrated near $n_B \cong n_F$ with a fluctuation of the order of $\sqrt{n_F}$ (Fig. 4). That is, for given n_F, the charge multiplicity distribution in the backward hemisphere would be stochastic.

Accepting this conclusion, we considered it[15] most natural that the energy and momentum distributions of the outgoing central region particles in the backward hemisphere, for this given n_F, would also be stochastic. It would be difficult to concoct a mechanism that could single out multiplicity distribution as the only one that is stochastic.

But the energy-momentum distribution must be subject to certain constraints:

(i) The total energy of the central particles is $E_0 h$ in each hemisphere. While the total energy in the backward direction is E_0 ($= \sqrt{s}/2 =$ the c.m. energy of the incoming hadron), only a fraction h of this would be carried by the central region particles in the backward hemisphere. The remaining fraction $1 - h$ would be taken away by the fragmentation (leading) particles in the backward direction. h depends on n_F.

(ii) Each emitted particle carries a bremsstrahlung (Bloch-Nordsieck) factor d^3p/E, which was first used[16] in multihadron production in the 1940's.

(iii) For each particle, there is a transverse-momentum cutoff factor which we take from experimental results to be

$$g(p_T) = \exp(-\alpha p_T) \tag{1.22}$$

where

$$\alpha = 5.25\ (\text{GeV/c})^{-1} \text{ for } 540 \text{ GeV } \bar{p}p \text{ collisions}. \tag{1.23}$$

(iv) Correlation effects among particles.

The simplest stochastic distribution of the particles in the backward hemisphere, taking into account constraints (i), (ii) and (iii), but not (iv), gives a probability factor of the form

$$\delta\left(\sum_i E_i - E_0 h\right) \prod_i (d^3 p_i/E_i) g(p_{Ti}) \tag{1.24}$$

where the summation and product extend over all the particles, charged or neutral, in the central region in the backward hemisphere. The single-particle spectrum can now be evaluated[16] from (1.24) by the method of steepest descent. For large multiplicities, this procedure gives a single-particle distribution

$$(d^3p/E) g(p_T) \exp(-E/T_p) \tag{1.25}$$

where T_p is a parameter which we have called the partition temperature. (For an alternative short cut derivation of this result, see Sec. 1.5.1 below.)

Equation (1.25) gives the single-particle distribution along the vertical line, at a fixed n_F, in Fig. 4. Since the event-populated region is assumed to be very narrow, as Fig. 4 indicates, fixing n_F is approximately equivalent to fixing n $(= n_F + n_B)$. Thus, we arrive at the conclusion that for each n there is a partition temperature T_p.

1.4.2. *Comparison with experimental angular distribution*

At 540 GeV for $\bar{p}p$ collisions the n_F–n_B distribution has not yet reached a narrow band. Nevertheless, we went ahead[15] to test the validity of (1.25) by evaluating from it the single-particle angular distribution. We write for the pseudorapidity η: ($\theta \equiv$ production angle in c.m. system)

$$\eta = \cosh^{-1}(1/\sin\theta). \tag{1.26}$$

Now,

$$dn/d\eta = 2\pi \sin^2\theta (dn/d\Omega). \tag{1.27}$$

Thus

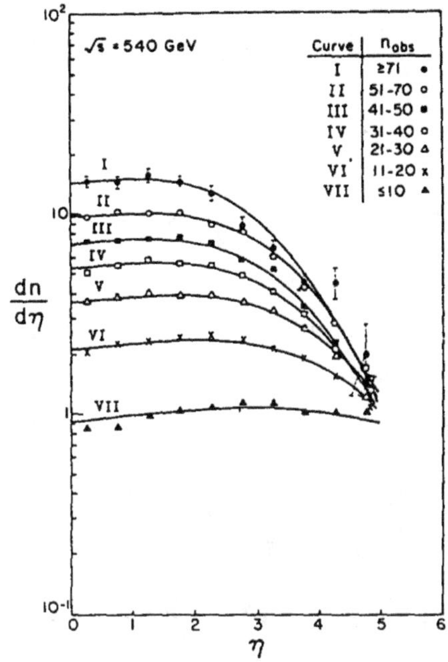

Fig. 5. $dn/d\eta$ versus η at $\sqrt{s} = 540$ GeV. Data points are taken from UA5 experiments. (Reproduced from Ref. 15.)

$$dn/d\eta = K2\pi \sin^2\theta \int_0^{E_0 h} p^2(dp/E)g(p\sin\theta)\exp(-E/T_p) \tag{1.28}$$

where K is a normalization constant. Only pions are included in this calculation. (Since the kaon to pion ratio increases with energy, for any accurate fit with experiments, the kaons will have to be included.) It is found that the curve for each multiplicity is well described by (1.28) for one value of T_p. Fig. 5 and Table 1 summarize these calculations. The good fit to the UA5 data indicates the usefulness of the idea of the partition temperature. It should be emphasized here that there are no adjustable parameters in this computation, the p_T-cutoff factor (1.22) having been taken from experiments. The energy fraction h and normalization constant K are both determined from the curves themselves. If one takes a Gaussian p_T distribution instead of (1.22), the fit to the angular distribution is also good.

1.5. *The concept of partition temperature*

1.5.1. *Origin of T_p*

The concept of T_p originates in

(i) the δ-function in (1.24) which controls the partition of energy on one side of the

Table 1. Parameters for multiparticle production processes at $\sqrt{s} = 540$ GeV. n_{obs}, which labels different charge-multiplicity ranges, is different from the true charge multiplicity by a factor of approximately 1.25 due to experimental corrections. For a given charge multiplicity, the shape of the experimental $dn/d\eta$ curve is well fitted by Eq. (1.28) for one value of T_p. The parameter h and normalization constant K are both determined from the curves themselves. The average values for the impact parameter are rough estimates based on a procedure outlined in T. T. Chou and C. N. Yang, *Phys. Lett.* **116B** (1982) 301. The total longitudinal momentum transfer is estimated by summing $(e - p_{\parallel})$ over all central-region particles (charged and neutral) using Eq. (1.28). Contributions to longitudinal momentum transfer from particles in the fragmentation region are negligible.

n_{obs}	Partition temperature T_p(GeV)	Normalization constant K(GeV^{-2})	n_{cal}	KNO variable $Z = n_{cal}/\bar{n}$	Energy fraction in central region $= h$	Average energy per particle in central region (GeV)	Average impact parameter b(fm) (approximate)	Longitudinal momentum transfer (GeV)
≥71	4.4	83	99	3.4	0.45	1.6	0.02	9.2
51–70	6.3	54	73	2.5	0.42	2.1	0.1	6.4
41–50	6.8	40	55	1.9	0.33	2.2	0.3	4.7
31–40	8.8	29	44	1.5	0.32	2.6	0.6	3.6
21–30	14.	20	33	1.1	0.31	3.4	0.9	2.5
11–20	24.	11	21	0.73	0.26	4.4	1.3	1.5
≤10	180.	4.8	11	0.37	0.20	6.6	1.8	0.67

collision, and

(ii) the method of steepest descent. The *mathematical procedure* used is exactly the same as that for obtaining the canonical ensemble from the microcanonical ensemble in statistical physics as the example below[17] illustrates.

Consider an ideal gas in a three-dimensional (x, y, z) region, $x > 0$, $y > 0$, $z > 0$, bounded by surfaces

$$x = 0, \quad y = 0, \quad z = 0, \quad \text{and}$$

$$z = yg(y)(m^2 + x^2 + y^2)^{-1/2} \tag{1.29}$$

where $g(y) > 0$, and m is a constant. The gas atoms are assumed to have infinite mass M (hence their kinetic energies are negligible,) moving in this region with a potential energy for each particle of

$$V = (m^2 + x^2 + y^2)^{1/2}. \tag{1.30}$$

A microcanonical ensemble for such a gas of N atoms at total energy E_0 has a probability distribution of

$$\delta\left[\sum_i^N V_i - E_0\right]\prod_i^N dx_i\, dy_i\, dz_i. \tag{1.31}$$

Integrating over all z_i gives

$$\delta\left[\sum_i (m^2 + x_i^2 + y_i^2)^{1/2} - E_0\right]\prod_i \frac{y_i}{(m^2 + x_i^2 + y_i^2)^{1/2}} g(y_i)\, dx_i\, dy_i. \tag{1.32}$$

If we make the replacements

$$\begin{aligned} x_i &\to (p_L)_i \\ y_i &\to (p_T)_i \\ E_0 &\to E_0 h, \end{aligned} \tag{1.33}$$

(1.32) becomes (1.24).

But this problem has a well known canonical ensemble with the single particle distribution

$$\exp(-V/\beta)\, dx\, dy\, dz,$$

or

$$\frac{y}{(m^2 + x^2 + y^2)^{1/2}} g(y) \exp[-(m^2 + x^2 + y^2)^{1/2}(1/\beta)]\, dx\, dy \tag{1.34}$$

after the coordinate z is integrated over, where β is the thermodynamical temperature of the gas. Substitution of (1.33) together with the replacement $\beta \to T_p$ converts (1.34) into (1.25).

We note that the number of parameters appearing in (1.25) and (1.34) depends on the number of δ-functions, i.e., constraints or conservation laws, in the problem. Consider the examples: In Planck's black-body radiation formula, there is only one parameter (T), which arises from the conservation of energy. The distribution of a free boson gas has two parameters corresponding to the conservation of energy and of the number of bosons in the system.

1.5.2. *Existence of T_p does not imply equilibrium*

Although for the gas model the temperature β is an equilibrium concept, for the high-energy collision problem T_p is just a mathematical parameter that governs the partition of energy in the stochastic process at impact parameter b and neither requires nor implies equilibrium. This point is particularly clear if we concentrate on the factor d^3p/E in (1.25). We want to make[17] the following observations.

(i) Replacing d^3p/E with d^3p leads to a single-particle distribution that totally disagrees with the angular distribution given by the UA5 group.

(ii) In any system in thermal equilibrium, such as for the black-body radiation, it is always the factor d^3p, not d^3p/E, that appears as the density of modes. Does the factor d^3p/E never occur for the problem of black-body radiation? The answer is, yes, it does occur when we discuss certain nonequilibrium phenomena. For example, consider in a block of matter at a finite temperature a cavity which at $t = 0$ is free of atoms and of all radiation. Immediately afterwards, short- and long-wave radiation would begin to fill the cavity. Since the coupling of radiation with matter contains the factor d^3p/E, the long wavelength modes generally couple more strongly with matter and are filled faster. Thus, in such a nonequilibrium situation, the factor d^3p/E does play a role. But in the long run, when equilibrium is established, *the strength of coupling is immaterial*, and only the mode density d^3p would play a role.

(iii) In a high-energy collision, there is not sufficient time to reach thermodynamical equilibrium in any overlapping part of the colliding hadrons. The Block-Nordsieck factor then plays a role and gives rise to the factor d^3p/E.

1.6. *Discussion of results*

1.6.1. *Increase of T_p with b*

Table 1 shows that the partition temperature increases with impact distance. This characteristic has a simple qualitative physical interpretation based on the geometrical picture. For a collision with a large impact parameter, the incoming hadrons have only a small overlap and the overlapping regions do not have much chance of exchanging longitudinal momenta. Therefore, they tend to maintain their respective original velocities in the center-of-mass system and few particles are emitted in the process. One thus expects the average energy per particle that results from these collisions to be large. That is, T_p should be large. On the other hand, for small impact parameters b, the two overlapping parts of the colliding hadrons are expected to exchange more longitudinal momenta, resulting in the emission of many particles with smaller average energy per particle, and therefore, smaller T_p.

1.6.2. *Inelasticity*

We assumed that the leading particles take away a fraction $1 - h$ of the total energy. This fraction according to Table 1 is larger than 50% for all collisions at 540 GeV. Averaged over all nondiffractive collisions, this fraction is $1 - \bar{h} = 72\%$. The remaining fraction $\bar{h} = 28\%$ of the incoming energy represents the average energy that goes into the production of central particles. This is what has been called "inelasticity" by cosmic-ray physicists.[4]

Experimental determination of the value of h for each collision has always in the past come up with an ambiguity, because it is not clear which of the fast particles in the c.m. system should be included in the "central region" and which in the fragmentation region. Since a fast particle contributes a large chunk of energy, this ambiguity has made it very difficult to give any accurate evaluation of the fraction h. With the introduction of the partition temperature T_p and (1.25), this difficulty disappears: Take the 540-GeV $\bar{p}p$ collision. For each multiplicity, T_p is determined from the experimental η distribution, mainly for low values of η, say $0 < \eta < 4$. Having determined T_p, (1.25) then allows for a calculation of the small tail of the curve for $\eta > 5$. The magnitude of this tail then makes possible an accurate evaluation of the probability of emission of a particle with $\eta > 5$ and, thereby, the contribution to h from the $\eta > 5$ part. Thus, this contribution is evaluated not from $\eta > 5$ data which are inaccurate, but from lower-η data together with (1.25).

1.6.3. *Longitudinal momentum transfer*

The value of the total longitudinal momentum transfer for 540-GeV $\bar{p}p$ collisions is listed in the last column of Table 1. We see that its value is very small for low-multiplicity events. Even for the high-multiplicity events, its value is only 9.2 GeV/c = 3.4% of the incoming momentum. This presents a picture for small-impact parameter collisions which is not what one would at first anticipate: For such collisions, the incoming hadrons are quite opaque to each other. The survival amplitude $S(b = 0) = \exp[-\Omega(0)]$ is[18] only ~ 0.19, so that only $\sim 4\%$ of such collisions are elastic. At first sight, this may suggest strong longitudinal momentum transfers for the remaining 96% of the collisions. But in fact, these remaining collisions are very far from a "bang", (i.e., an amalgamation of the two hadrons in one region of space, as in the early Fermi[19] picture). Instead, the longitudinal momentum exchange between the two sides is only $\sim 3\%$. (Of course, there is bound to be a process in which thousands of particles are emitted in the c.m. system, all nearly at rest. Such processes would involve a large longitudinal momentum transfer of ~ 270 GeV/c, and would require $h = 1$. But that would clearly have an extremely small cross section.)

The smallness of the longitudinal momentum transfer, even for processes involving the production of many particles, indicates what we may call the *persistence of longitudinal momentum* in very-high-energy hadron-hadron collisions. This is a dominating characteristic: For the 540-GeV $\bar{p}p$ collisions, elastic and double-diffraction-dissociation events involve longitudinal momentum transfers of the order of at most a dozen MeV's, while the nondiffractive events involve longitudinal momentum transfers of the order of a few GeV's, not more. Larger longitudinal momentum transfers are *extremely* rare.

1.6.4. *Inclusion of particles other than pions*

We speculate that the observed π/K ratio in outgoing particles in $\bar{p}p$ collisions could be just the result of the effect of the masses m_π and m_K in (1.25), when the incoming energy is high enough.

1.7. *Further predictions*

We present here some additional predictions which are not included in the above discussions.

1.7.1. *Spin-spin correlation of emitted particles*

The large angular momentum involved in a high energy collision[20] would lead to transverse-spin-transverse-spin correlations for the outgoing particles which we shall explain as follows with a concrete example.

Consider a $\bar{p}p$ collision at c.m. energy of 540 GeV, producing ~21–30 observed charge particles. The average impact parameter[17] involved is ~0.9 fm so that the incoming total angular momentum is ~$1000\hbar$. The total number of outgoing hadrons[17] is ~$33 \times (3/2) = 49$. How do these 49 particles combine to carry away 1000 units of angular momentum?

Much of it is certainly carried away as orbital angular momenta by the outgoing particles. But the sum of all these orbital angular momenta *is not sufficient* to make up the original angular momentum, as the following analysis will demonstrate: Consider all the outgoing pieces that move toward the right in Fig. 6, including the leading particles and those in the central region. Let b_i and p_i^L be the impact parameter[21] and the longitudinal momentum of the ith such piece. (b_i may be positive or negative according as whether the point of origin P in Fig. 6 is above or below the main axis.)

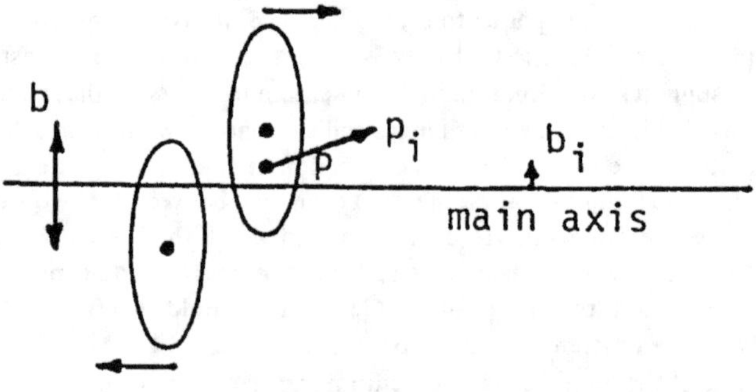

Fig. 6. Impact parameter b_i and momentum p_i of outgoing particle i originating at point P. (Reproduced from Ref. 20.)

It contributes an orbital angular momentum of $b_i p_i^L$. The total orbital angular momentum of all right-moving outgoing particles is then

$$\sum_R b_i p_i^L = \sum_R b_i E_i - \sum_R b_i (E_i - p_i^L). \tag{1.35}$$

Now there is little energy transfer in the collision. So E_i is also the energy of the piece of the incoming hadron at point P, which is the same as the longitudinal momentum, $(p_i^L)^{\text{in}}$, of that piece of the incoming hadron. Thus,

$$\sum_R b_i p_i^L = \sum_R b_i (p_i^L)^{\text{in}} - \sum_R b_i [(p_i^L)^{\text{in}} - p_i^L]. \tag{1.36}$$

Now $\sum_R b_i (p_i^L)^{\text{in}}$ is the incoming right-moving hadron's contribution to the incoming angular momentum. Thus, there is a net deficiency of orbital angular momentum in the right-moving outgoing system, by an amount

$$\sum_R b_i [(p_i^L)^{\text{in}} - p_i^L]. \tag{1.37}$$

To estimate the magnitude of this deficiency we equate it approximately with

$$\bar{b}_i \sum_R [(p_i^L)^{\text{in}} - p_i^L]. \tag{1.38}$$

Now $\bar{b}_i$ is approximately[17] (1/2)(0.9 fm). The sum in (1.38) is the longitudinal momentum transfer which was estimated[17] to be 2.5 GeV/c. Thus, the orbital angular momentum deficiency for the right moving particles is $\sim(1/2)(0.9\text{ fm})(2.5\text{ GeV/c}) \cong 6\hbar$.

In other words, the 49 outgoing particles must contribute $\sim 12\hbar$ of spin angular momentum in order to make up the total angular momentum. There is thus a strong necessity for the spins of the outgoing particles to line up *parallel* to each other in the transverse direction:

$$(\sigma_1^T \cdot \sigma_2^T)_{\text{average}} > 0. \tag{1.39}$$

Notice that in this prediction, particles 1 and 2 may be both right moving, left moving or they may be moving towards different sides. Prediction (1.39) should not be difficult to check for outgoing Λ's and Σ's.

1.7.2. *Decrease of single-diffraction events*

If for higher and higher energies, the $n_F - n_B$ multiplicity distribution becomes a narrow band as predicted, strongly forward-backward asymmetrical events will become progressively unimportant. In particular, the fraction of single-diffraction events, events populating near the n_F- and n_B-axis in Fig. 1 for which there are only a few diffractive fast particles on one side and many particles emitted on the other, will become rare. This is consistent with the recent UA4 observation[22] that the ratio of single-diffraction to total cross sections decreases with increasing energy.

1.8. *Extrapolation to higher and lower energies*

We have made[17] extrapolations of the angular distribution to $\sqrt{s} = 53$ GeV (CERN-ISR), 2 TeV (Fermilab Tevatron) and 40 TeV (SSC) with the following assumptions:

(i) The p_T cutoff factor $g(p_T)$ is taken to be $g(p_T) = \exp(-\alpha p_T)$, where we take

$$\alpha = 5.8\ (\text{GeV/c})^{-1} \qquad \text{for } \sqrt{s} = 53\ \text{GeV},$$

$$\alpha = 5.0\ (\text{GeV/c})^{-1} \qquad \text{for } \sqrt{s} = 2\ \text{TeV},$$

$$\alpha = 4.4\ (\text{GeV/c})^{-1} \qquad \text{for } \sqrt{s} = 40\ \text{TeV}.$$

(ii) The value of $\bar{n}_{\text{ch}}$ for these energies are taken to be 13, 41 and 78, respectively.
(iii) The parameter h is taken to be a function only of the KNO variable $Z = n/\bar{n}$.

The results are presented in Fig. 7 and Table 2.

2. Multiparticle Production in Two-Jet Events in e^+e^- Collision

2.1. *Does e^+e^- collision give stochastic (i.e., narrow) or nonstochastic (i.e., wide) multiplicity distribution?*

The crucial conclusion of Sec. 1 above is that the wide multiplicity distribution in $\bar{p}p$ collisions is due to the large fluctuation of the impact parameter, or angular momentum, of the collision. If this conclusion is correct, then it follows that for e^+e^- collision, where the total angular momentum is 0 or 1, (which is the angular momentum of the virtual intermediate photon), *there should be no wide multiplicity fluctuation.* In fact, the natural assumption is that the multiplicity distribution in e^+e^- collisions is a Poisson distribution. For those events which are classified as two-jet events, the natural assumption is that the distribution $\mathbb{P}(n_F, n_B)$ is a product of two Poissons, one in n_F, one in n_B. These assumptions imply that the multiplicity distributions are stochastic (narrow). We made[14,23] these conjectures in 1985 in an effort to gain a unified understanding of multiplicity distributions in $\bar{p}p$ and e^+e^- collisions. In particular, we conjectured the following:

(i) The distribution of the forward, backward charged multiplicities $\mathbb{P}(n_F, n_B)$ is a product distribution:

$$\mathbb{P}(n_F, n_B) = g(n_F)h(n_B). \tag{2.1}$$

(ii) The separate distribution on each side is Poisson with a "cluster" size k:

$$\mathbb{P}(n_F, n_B) = f(a, n_F/k)f(a, n_B/k) \tag{2.2}$$

where

$$f(a, n_F/k) = (\text{constant})\ a^{(n_F/k)}/(n_F/k)! \tag{2.3}$$

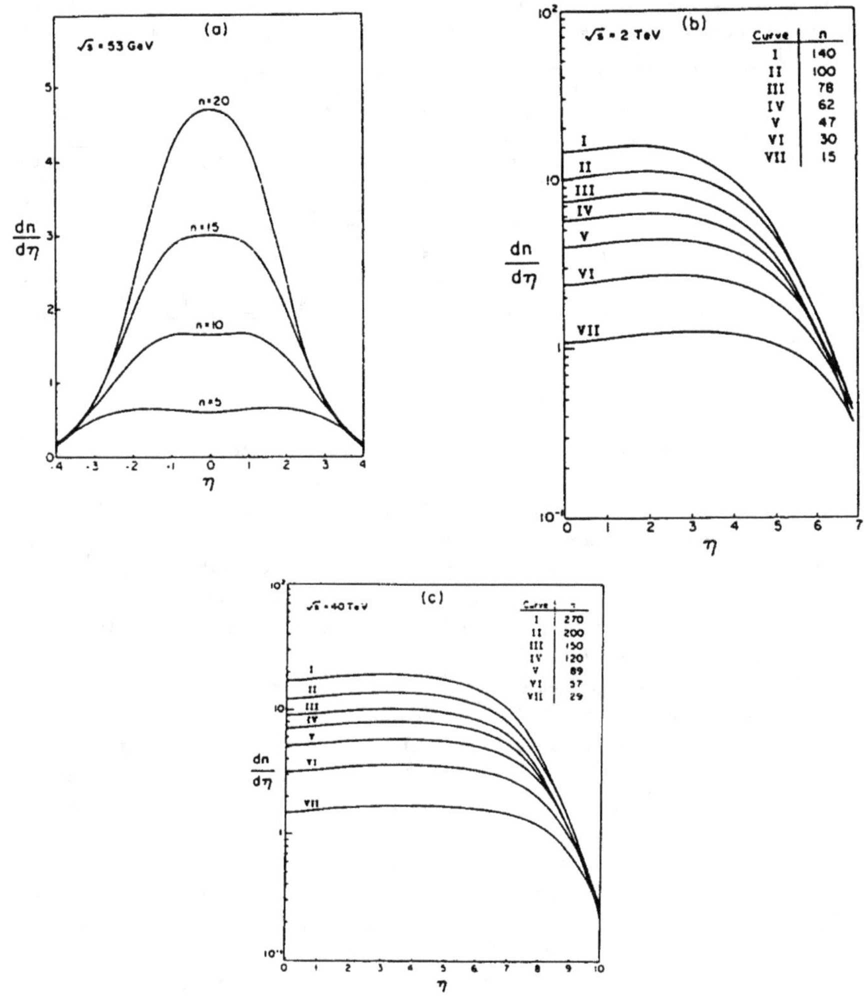

Fig. 7. Calculated $dn/d\eta$ versus η at (a) $\sqrt{s} = 53$ GeV, (b) $\sqrt{s} = 2$ TeV, and (c) $\sqrt{s} = 40$ TeV. (Reproduced from Ref. 17.)

(iii) We conjectured further that $k = 2$ in analogy with the distribution[8,9] with respect to $z = n_F - n_B$ for $\bar{p}p$ collisions.

2.2. *Comparison with experiments*

2.2.1. *Multiplicity distribution curves*

The most important qualitative difference between a KNO or negative binomial distribution and a Poisson distribution is

Table 2. Parameters for single-particle distribution at ISR, Tevatron, and SSC energies. The relation between α and $\langle p_T \rangle$ is given by $\alpha = 2\langle p_T \rangle^{-1}$. n is the total charge multiplicity. The normalization constant K is in GeV^{-2} and partition temperature T_p in GeV.

		$\sqrt{s} = 53$ GeV ($\alpha = 5.8\ \mathrm{GeV}^{-1}$)			$\sqrt{s} = 2$ TeV ($\alpha = 5.0\ \mathrm{GeV}^{-1}$)			$\sqrt{s} = 40$ TeV ($\alpha = 4.4\ \mathrm{GeV}^{-1}$)		
Z	$h(Z)$	n	K	T_p	n	K	T_p	n	K	T_p
3.4	0.45	44	350	0.25	140	71	17	270	62	300
2.5	0.42	33	140	0.43	100	48	23	200	44	400
1.9	0.33	25	91	0.49	78	36	25	150	33	420
1.5	0.32	20	53	0.70	62	27	31	120	26	520
1.2	0.31	15	27	1.2	47	19	43	89	18	700
0.73	0.26	10	13	2.0	30	11	61	57	11	950
0.37	0.20	5	3.9	32.	15	5.1	130	29	5.3	1700

$$\Delta n/\bar{n} = (\text{constant})(\bar{n})^{-1/2} \qquad (\text{Poisson}) \tag{2.4}$$

$$\Delta n/\bar{n} > (\text{constant}) \qquad (\text{KNO or negative binomial}). \tag{2.5}$$

This difference in principle could resolve the question whether in e^+e^- collisions a Poisson distribution is obtained. Unfortunately, existing data for e^+e^- collisions cover too narrow a range of values for $\bar{n}$ and are insufficient to resolve this question.

But the shapes of the multiplicity distribution for $\bar{p}p$ and e^+e^- collisions are distinctively different (Fig. 8) with the latter case fitting Poisson distribution well. In the recent experiment of Derrick *et al.*[24] at 29 GeV for e^+e^- collisions, the fit with a Poisson is even better (Fig. 9): We emphasize that with a Poisson distribution, there is to the left of the peak a region of the curve where the second derivative is positive. As $\bar{n}$ increases, this region characteristically widens. Both of these phenomena are apparent in the e^+e^- curves in Fig. 8, but are absent in the $\bar{p}p$ curves. Furthermore, e^+e^- curves are quite symmetrical on the two sides of the peak, which is a characteristic also of Poisson distributions but not of the $\bar{p}p$ curves.

2.2.2. *Consequence of recent results of Derrick et al.*

We shall demonstrate[25] below that the recent careful analysis of Derrick *et al.*[24] of their e^+e^- 29-GeV data strongly indicates that conjectures (i) and (ii) above are correct; but conjecture (iii) is in error, since their result indicates that $k = 1$ instead of $k = 2$.

Derrick *et al.*[24] found that for fixed n_F, the average $\langle n_B \rangle$ is essentially independent of n_F. The average slope was given as

$$b = -0.001 \pm 0.015 \tag{2.6}$$

which is also consistent with earlier TASSO data.[26] Although these results do not prove the product form (2.1) of the n_F–n_B distribution, it is strongly suggestive of it. Assuming (2.1) we use the condition

$$\mathbb{P}(n_F, n_B) = \mathbb{P}(n_B, n_F)$$

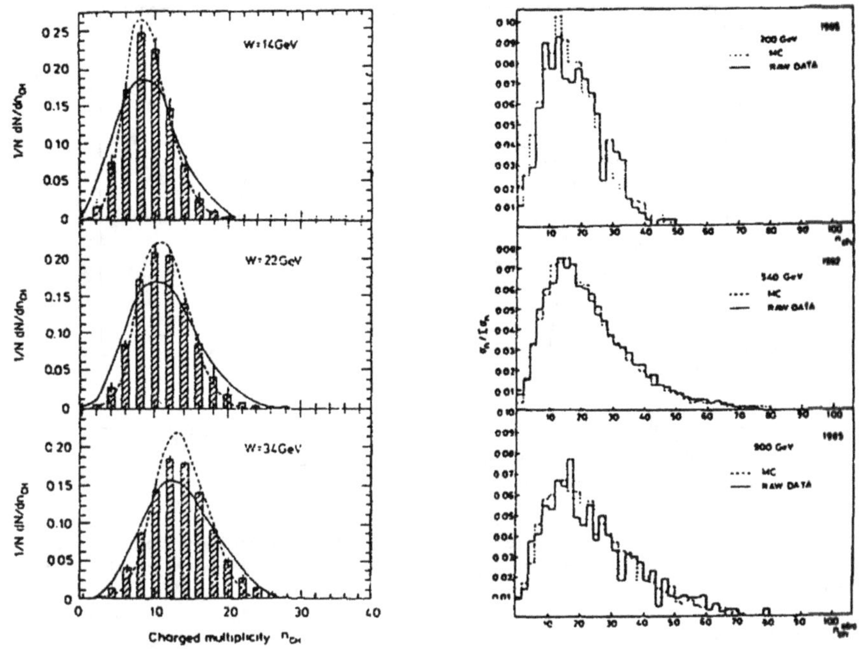

Fig. 8. Comparison of charged-particle multiplicity distribution for e^+e^- annihilation and $\bar{p}p$ collisions. The e^+e^- curves (left column) are reproduced from a TASSO publication, Ref. 27. No corrections for gluon bremsstrahlung effects have been made. The $\bar{p}p$ curves (right column) are reproduced from a UA5 report. Ref. 12. For details see these original papers.

which follows from charge conjugation invariance of the strong and electromagnetic interactions. Thus

$$g(x)h(y) = g(y)h(x)$$

i.e.,

$$g(x)/h(x) = g(y)/h(y).$$

Since x and y are independent, this shows that $g(x)/h(x) =$ constant. By normalizing g and h, we see that they are identical and we arrive that

$$\mathbb{P}(n_F, n_B) = g(n_F)g(n_B). \tag{2.7}$$

Now Derrick *et al.*[24] found that the distribution of $n = n_F + n_B$ fits a Poisson beauti-

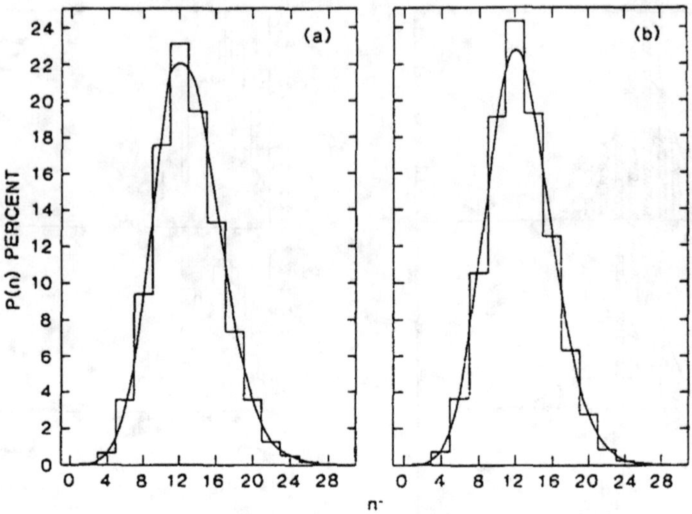

Fig. 9. Charged-particle multiplicity measurements (histogram) compared to the line that connects points on a Poisson distribution with the same mean value: (a) inclusive data sample, (b) two-jet data sample. (Reproduced from Ref. 24.)

fully. In the language of (2.7), this means

$$\sum_{n_F} g(n_F)g(n-n_F) = (\text{constant})\ a^n/n! \qquad (n = \text{even}). \tag{2.8}$$

Equation (2.8) can be solved with the method of generating functions as follows. Define

$$G(x) \equiv \sum_{m=0}^{\infty} g(m)x^m. \tag{2.9}$$

Then

$$[G(x)]^2 = \sum_{m_1,m_2} x^{m_1+m_2} g(m_1)g(m_2)$$

which on account of (2.8) becomes

$$[G(x)]^2 = (\text{constant}) \sum a^n x^n/n! = (\text{constant})\ e^{ax}. \tag{2.10}$$

Thus

$$G(x) = (\text{constant})\, e^{ax/2}$$

which leads to the explicit expression for $g(m)$ as a Poisson

$$g(m) = (\text{constant})\, (a/2)^m/m!. \tag{2.11}$$

Comparison with (2.3) shows that the cluster size is $k = 1$ (identifying $n_F = m$), and not $k = 2$. Thus, we conclude that the results of Derrick *et al.*[24] strongly suggest the validity of conjectures (i) and (ii), but give $k = 1$ instead of $k = 2$.

The above derivation is not exact because $n = n_F + n_B$ must be even. Thus, the n_F, n_B values form a checkerboard pattern and (2.7) is an interpolative equation. In such an interpolation, the value of $\sum_{n_F} g(n_F)g(n - n_F)$ for $n =$ odd is also approximately given by (2.8). Thus, (2.10) is approximately, but not exactly correct. Therefore, (2.11) is also approximately but not exactly correct.

Confirmation of conjectures (i) and (ii) gives strong support to our unified physical picture[14,23] that the multiplicity distribution of e^+e^- is narrow and that of $\bar{p}p$ is broad because of the wide range of angular momentum available in the latter case. But why should the cluster size be $k = 2$ for $\bar{p}p$ and $k = 1$ for e^+e^-? We believe the answer perhaps resides in the same mechanism that produces[14] another difference between the two cases: The different average charged multiplicities when one concentrates on those events with zero impact parameter in $\bar{p}p$ collisions and compares them with two-jet e^+e^- events at a proper energy (Fig. 10). The quark-gluon system in the two

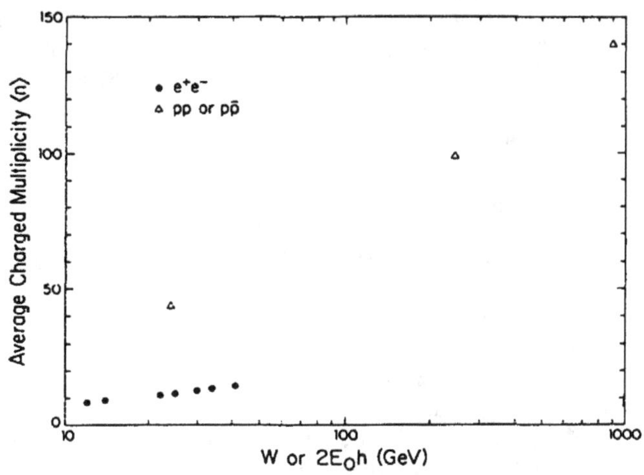

Fig. 10. Plot of average charged multiplicity versus total energy W for e^+e^- annihilation or central energy $2E_0h$ for $\bar{p}p$ collisions at approximately zero impact parameter. The average multiplicities for $\bar{p}p$ collisions are crude estimates subject to large uncertainties. (Reproduced from Ref. 14.)

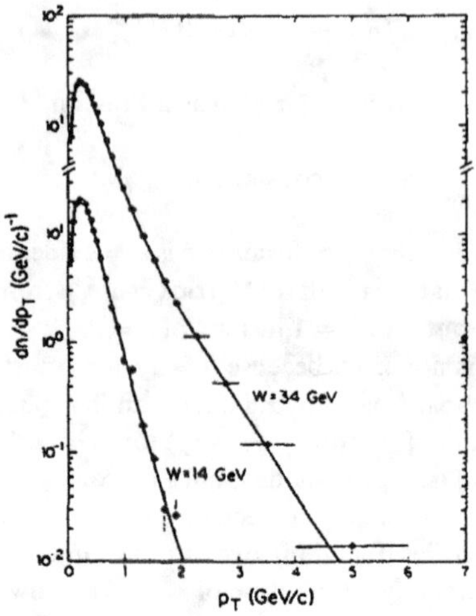

Fig. 11. The calculated dn/dp_T distributions for e^+e^- annihilation at $W = 14$ and 34 GeV. The experimental data points are taken from Ref. 27. (Reproduced from Ref. 14.)

cases start from very different[14] initial conditions, even though with the same total energy. The resultant average cluster size and average multiplicity may, therefore, be different.

2.2.3. *Partition temperature T_p*

The same argument that we gave in Sec. 1.4 for the single-particle momentum distribution in $\bar{p}p$ collisions applies also to e^+e^- collisions, the only difference being that for e^+e^- collisions there is only one partition temperature because there is only one impact parameter $b = 0$ for a given $W = \sqrt{s}$. We have fitted[14] the TASSO data[27] for the momentum distribution of single charged particles and found that

$$\begin{aligned} &\text{at } W = 14\text{ GeV} \qquad T_p = 1.6\text{ GeV},\\ &\text{at } W = 34\text{ GeV} \qquad T_p = 3.3\text{ GeV}. \end{aligned} \tag{2.12}$$

The fits are good and are reproduced in Figs. 11 and 12.

2.2.4. *Root-mean-square net charge in each jet*

Although the mechanism for the different values of the cluster size, $k = 1$ for e^+e^- and $k = 2$ for $\bar{p}p$ collisions, is not clear, we observe they are based on the very good fits in Fig. 9 (for e^+e^-) and Fig. 3 (for $\bar{p}p$), both of which are parameterless (except for k). Without understanding the underlying mechanism, are there nevertheless other experimentally measureable quantities that may be correlated with the cluster size?

We believe that the answer is yes. The root-mean-square of the net charge Q in each jet is such a quantity. To see this, we observe that a natural mechanism for cluster size

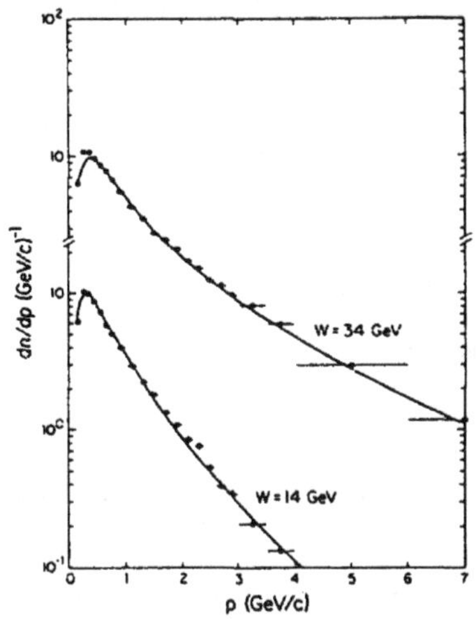

Fig. 12. The calculated charged-particle momentum spectrum dn/dp for e^+e^- annihilation at $W = 14$ and 34 GeV. The experimental data points are taken from Ref. 27. (Reproduced from Ref. 14.)

$k = 2$ is that neutral clusters are emitted stochastically, each of which then splitting into one positively and one negatively charged particle. There must not be too much cross-overs in this splitting process. (A cross-over is one where the splitting puts the positively and negatively charged particles into the two different hemispheres, forward and backward.) Otherwise, the observed $n_F - n_B$ distribution would not be like (1.7) which is obtained when there are no cross-overs at all.

If there are no cross-overs, then the net charge in each hemisphere should be essentially zero:

$$\overline{Q^2} \cong 0. \tag{2.13}$$

On the other hand, for cluster size $k = 1$, the natural mechanism is that the particles are stochastically emitted, so that there is equal forward and backward probability for each outgoing particle. By the central limit theorem the root-mean-square net charge in each hemisphere should then be

$$(\overline{Q^2})^{1/2} = (\text{constant})\, n^{1/2}. \tag{2.14}$$

We have studied[23] a model like this and found

$$(\overline{Q^2})^{1/2} = (1/2) n^{1/2}. \tag{2.15}$$

The details are as follows:

Let the number of particles with charge $+1$ or -1 be both equal to $n/2$ and let each particle be randomly assigned to either the forward or backward hemisphere. We can find the distribution of the net charge Q in the forward hemisphere by the method of generating functions: The probability that there are Q_+ particles of charge $+1$ in the forward hemisphere is the coefficient of x^{Q_+} in

$$[(1/2)(1+x)]^{n/2}.$$

Similarly, the probability that there are Q_- particles of charge -1 in the forward hemisphere is the coefficient of $x^{-(Q_-)}$ in

$$[(1/2)(1+x^{-1})]^{n/2}.$$

The net charge in the forward hemisphere $Q = Q_+ - Q_-$. Thus, the probability for Q is the coefficient of x^Q in

$$[(1/2)(1+x)(1/2)(1+x^{-1})]^{n/2}. \tag{2.16}$$

It is now easy to evaluate for this model the moments of Q

$$\bar{Q} = 0, \qquad \overline{Q^2} = n/4, \tag{2.17}$$

i.e.,

$$(\overline{Q^2})^{1/2} = (1/2)n^{1/2}. \tag{2.15}$$

There is not much data with which to check (2.14). TASSO[28] has reported that $\overline{Q^2} = 1.53$ at $W = 33$ GeV, at which $\bar{n} = 13$.

2.2.5. *Lack of $\sigma_1^T \cdot \sigma_2^T$ correlation*

The prediction about spin correlations in (1.39) for $\bar{p}p$ collisions was based on the large average transverse angular momentum of $\bar{p}p$ events. Since the average transverse angular momentum is small for e^+e^-, we predict that for e^+e^- two-jet events

$$(\sigma_1^T \cdot \sigma_2^T)_{\text{average}} = 0. \tag{2.18}$$

Acknowledgments

This work is supported in part by the U. S. Department of Energy under grant No. DE-FG09-84ER40160 and by the National Science Foundation under grant No. PHY 85-07627.

References

1. T. T. Chou and C. N. Yang, in *High-Energy Physics and Nuclear Structure*, ed. G. Alexander (North-Holland, Amsterdam, 1967), pp. 348–359; *Phys. Rev.* **170** (1968) 1591; *Phys. Rev. Lett.* **20** (1968) 1213; *Phys. Rev.* **175** (1968) 1832.
2. Further theoretical work on elastic scattering can be found in A. W. Chao and C. N. Yang, *Phys. Rev.* **D8** (1973) 2063; F. Hayot and U. P. Sukhatme, *Phys. Rev.* **D10** (1974) 2183; T.

T. Chou, *Phys. Rev.* **D11** (1975) 3145 and **D19** (1979) 3327; T. T. Chou and C. N. Yang, *Phys. Rev.* **D17** (1978) 1889 and **D19** (1979) 3268; *Phys. Rev. Lett.* **46** (1981) 764 and 1350(E); T. T. Chou, T. T. Wu and C. N. Yang, *Phys. Rev.* **D26** (1982) 328; T. T. Chou and C. N. Yang, *Phys. Lett.* **128B** (1983) 457. For experimental support, see A. Bohm *et al., Phys. Lett.* **49B** (1974) 491; N. Kwak *et al., Phys. Lett.* **58B** (1975) 233; E. Nagy *et al., Nucl. Phys.* **B150** (1979) 221; UA4 Collaboration, M. Bozzo *et al., Phys. Lett.* **147B** (1984) 392; S. R. Amendolia *et al., Phys. Lett.* **146B** (1984) 116 and **178B** (1986) 435.

3. J. Benecke, T. T. Chou, C. N. Yang and E. Yen. *Phys. Rev.* **188** (1969) 2159; T. T. Chou and C. N. Yang, *Phys. Rev. Lett.* **25** (1970) 1072; T. T. Chou, *Phys. Rev. Lett.* **27** (1971) 1247; T. T. Chou and C. N. Yang, *Phys. Rev.* **D7** (1973) 1425; T. T. Chou and C. N. Yang, *Nucl. Phys.* **B107** (1976) 1; N. Byers and C. N. Yang, *Phys. Rev.* **142** (1966) 976; T. T. Chou, *Phys. Rev.* **176** (1968) 2041; T. T. Chou and C. N. Yang, *Phys. Rev.* **D22** (1980) 610.
4. "Leading particle" is a term used by cosmic ray physicists to characterize outgoing particles in the very forward and backward directions with large energies. If we write the total energy carried away by such particles divided by $\sqrt{s}$ as $1 - h$, then h was called the inelasticity. See G. Cocconi, *Phys. Rev.* **111** (1958) 1699. The nonleading particles are called central region particles.
5. UA5 Collaboration, K. Alpgård *et al., Phys. Lett.* **107B** (1981) 310 and 315; and UA1 Collaboration, G. Arnison *et al., Phys. Lett.* **107B** (1981) 320.
6. Z. Koba, H. B. Nielsen and P. Olesen, *Nucl. Phys.* **B40** (1972) 317.
7. UA5 Collaboration, G. J. Alner *et al., Phys. Lett.* **160B** (1985) 193 and 199. This is the latest experimental analysis. The introduction of negative binomial distributions into the analysis of multiplicity data involves many papers and authors. They will not be listed here.
8. UA5 Collaboration, K. Alpgård *et al., Phys. Lett.* **123B** (1983) 361.
9. T. T. Chou and C. N. Yang, *Phys. Lett.* **135B** (1984) 175.
10. A model which illustrates Eq. (1.9) can be found in T. T. Chou and C. N. Yang, *Phys. Lett.* **167B** (1986) 453.
11. UA5 Collaboration, G. J. Alner *et al., Nucl. Phys.* (to be published).
12. J. G. Rushbrooke, in *Multiparticle Dynamics 1985*, Proceedings of the XVI Symposium, Kiryat Anavim, Israel, 1985, ed. J. Grunhaus (World Scientific, Singapore, 1985), p. 289.
13. This assumption is very natural in the geometrical picture. It is discussed in the literature by S. Barshay, *Phys. Lett.* **42B** (1972) 457; H. B. Nielsen and P. Olesen, *Phys. Lett.* **43B** (1973) 37; A. Bialas and E. Bialas, *Acta Phys. Pol.* **B5** (1974) 373; S. Barshay and Y. Yamaguchi, *Phys. Lett.* **51B** (1974) 376; T. T. Chou and C. N. Yang, *Phys. Lett.* **116B** (1982) 301.
14. T. T. Chou and C. N. Yang, *Phys. Rev. Lett.* **55** (1985) 1359.
15. T. T. Chou, C. N. Yang and E. Yen, *Phys. Rev. Lett.* **54** (1985) 510.
16. F. Bloch and A. Nordsieck, *Phys. Rev.* **52** (1937) 54; H. W. Lewis, J. R. Oppenheimer and S. A. Wouthuysen, *Phys. Rev.* **73** (1948) 127.
17. T. T. Chou and C. N. Yang, *Phys. Rev.* **D32** (1985) 1692.
18. T. T. Chou and C. N. Yang, *Phys. Lett.* **128B** (1983) 457.
19. E. Fermi, *Prog. Theor. Phys.* **5** (1950) 570; *Phys. Rev.* **81** (1951) 683.
20. C. N. Yang and T. T. Chou, *J. Phys. Soc. Japan.* **55** (1986) 53 Suppl.
21. The impact parameter is a two-dimensional quantity. b_i is the projection of the impact parameter into the impact plane. Similarly for p_i^L.
22. UA4 Collaboration, D. Bernard *et al., Phys. Lett.* **186B** (1987) 227.
23. T. T. Chou and C. N. Yang, *Phys. Lett.* **167B** (1986) 453 and **171B** (1986) 486(E).
24. M. Derrick *et al., Phys. Rev.* **D34** (1986) 3304.
25. T. T. Chou and C. N. Yang, *Phys. Lett.* **193B** (1987) 531.
26. TASSO Collaboration, M. Althoff *et al., Z. Phys.* **C29** (1985) 347.
27. TASSO Collaboration, M. Althoff *et al., Z. Phys.* **C22** (1984) 307.
28. TASSO Collaboration, R. Brandelik *et al., Phys. Lett.* **100B** (1981) 357.

https://doi.org/10.1142/9789819826636_0026

SO_4 SYMMETRY IN A HUBBARD MODEL

CHEN NING YANG
Institute for Theoretical Physics, State University of New York, Stony Brook, NY 11794-3840, USA

and

S. C. ZHANG
IBM Research Division, Almaden Research Center, San Jose, CA 95120-6099, USA

Received 28 May 1990

For a simple Hubbard model, using a particle-particle pairing operator η and a particle-hole pairing operator ζ, it is shown that one can write down two commuting sets of angular momenta operators **J** and **J**$'$, both of which commute with the Hamiltonian. These considerations allow the introduction of quantum numbers j and j', and lead to the fact that the system has $SO_4 = (SU_2 \times SU_2)/Z_2$ symmetry. j is related to the existence of superconductivity for a state and j' to its magnetic properties.

In a recent paper[1] it was found that a pairing operator η is useful for considering the Hamiltonian in a simple Hubbard model on an $L \times L \times L$ lattice, where L = even. We shall extend such considerations in the present paper. All notations are the same as in Ref. 1. We introduce here a Hamiltonian H' and a momentum operator $\mathbf{P}'$ which are trivially different from the H and $\mathbf{P}$ of Ref. 1, in order to bring out more *symmetries* of the system:

$$H' = T' + V' \,, \tag{1}$$

$$T' = -2\varepsilon \sum_{\mathbf{k}} (\cos k_x + \cos k_y + \cos k_z)(a_{\mathbf{k}}^+ a_{\mathbf{k}} + b_{\mathbf{k}}^+ b_{\mathbf{k}}) \,, \tag{2}$$

$$V' = 2W \sum_{\mathbf{r}} \left(a_{\mathbf{r}}^+ a_{\mathbf{r}} - \frac{1}{2} \right)\left(b_{\mathbf{r}}^+ b_{\mathbf{r}} - \frac{1}{2} \right) . \tag{3}$$

$$\mathbf{P}' = \Sigma\left(\mathbf{k} - \frac{1}{2}\boldsymbol{\pi}\right)(a_{\mathbf{k}}^+ a_{\mathbf{k}} + b_{\mathbf{k}}^+ b_{\mathbf{k}})\,(\mathrm{mod.}2\pi) \,. \tag{4}$$

(1) *The operators J_x, J_y, and J_z* — It is easy to verify that $\eta^+\eta - \eta\eta^+ = \Sigma(a^+a + b^+b) - M$, where $M = L^3$. Calculating the commutator of this commutator with η we obtain

PACS Nos: 74.20.-z, 05.30.Fk.

Reproduced from International Journal of Modern Physics B, Vol. 05, No. 06n07, pp. 977–984 (1991), DOI: S021797929100050X

Theorem 1. Defining

$$\eta^+ = J_x + iJ_y\,,\quad \eta = J_x - iJ_y\,,\quad J_z = \frac{1}{2}\Sigma\,(a^+a + b^+b) - \frac{1}{2}M\,, \tag{5}$$

one finds that J_x, J_y, J_z commute with each other like the components of an angular momentum. Hence the eigenvalue of $\mathbf{J}^2$ is $j(j+1)$ where $2j$ = integer $\geqq 0$. Furthermore (as can be easily checked),

$$[T',\mathbf{J}]_- = [V',\mathbf{J}]_- = [H',\mathbf{J}]_- = [\mathbf{P}'\mathbf{J}]_- = 0\,. \tag{6}$$

(2) *The operators J'_x, J'_y and J'_z* — We now define a particle-hole pairing operator,

$$\zeta = \Sigma\, a_k b_k^+ = \Sigma\, a_r b_r^+\,. \tag{7}$$

Then

$$\zeta\zeta^+ - \zeta^+\zeta = -\Sigma\, a^+a + \Sigma\, b^+b\,.$$

Theorem 2. Defining

$$\zeta^+ = J'_x + iJ'_y\,,\quad \zeta = J'_x - iJ'_y\,,\quad J'_z = \frac{1}{2}\Sigma\, a^+a - \frac{1}{2}\Sigma\, b^+b\,, \tag{8}$$

one finds that J'_x, J'_y, J'_z commute with each other like the components of an angular momentum. Hence the eigenvalue of $\mathbf{J}'^2$ is $j'(j'+1)$ where $2j'$ = integer $\geqq 0$. Furthermore all 3 components of $\mathbf{J}$ commute with all 3 components of $\mathbf{J}'$, and

$$[T',\mathbf{J}']_- = [V',\mathbf{J}']_- = [H',\mathbf{J}']_- = [\mathbf{P}'\mathbf{J}']_- = 0\,. \tag{9}$$

ζ is the usual spin lowering operator and $\mathbf{J}'$ is the usual "spin" operator.

(3) *Explicit eigenfunctions of H'* — We can find many eigenstates of H' with Theorems 1 and 2 as follows. We diagonize $\mathbf{J}^2$, $\mathbf{J}'^2$, J_z, J'_z, H' and $\mathbf{P}'$ simultaneously. These states can be sorted out into multiplets $\{j, j'\}$, each comprising of $(2j+1)(2j'+1)$ states, as illustrated in Fig. 1, where N_a and N_b are eigenvalues of Σa^+a and Σb^+b,

$$j_z = \frac{1}{2}(N_a + N_b - M)\,,\quad j_z = \frac{1}{2}(N_a - N_b)\,. \tag{10}$$

As explained in Fig. 1, $j + j'$ = integer, i.e., not all representations of $SU_2 \times SU_2$ are present. This means that the true symmetry of the problem is $(SU_2 \times SU_2)/Z_2 = SO_4$.

Consider now the states in one spot on the bottom row of Fig. 1. For these states, $N_a = 0$. The operators H' and $\mathbf{P}'$ for such states are easily diagonizable since for such states, there are no a-particle — b-particle interac-

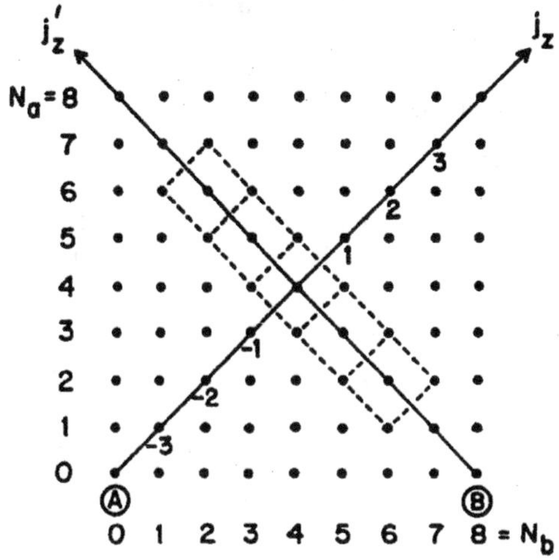

Fig. 1. (N_a, N_b) diagram for $M = 8$. The relationship between (j_z, j'_z) with (N_a, N_b) is given by Eq. (10). Each multiplet $\{j, j'\}$ is represented by a rectangular set of states centered at $j_z = j'_z = 0$ in this diagram. The number of states in the multiplet is $(2j+1)(2j'+1)$. Illustrated is the multiplet $\left\{\frac{1}{2}, \frac{5}{2}\right\}$. All states of a multiplet share the same eigenvalue of H' and $\mathbf{P}'$. The lowest corner in the multiplet is where $j_z = -j$, $j'_z = -j'$. One can generate all states of a multiplet by starting from its lowest corner and repeatedly operate on it with $\eta^+ = J_x + iJ_y$ (which increases j_z) and with $\zeta^+ = J'_x + iJ'_y$ (which increases j'_z). Obviously $j + j'$ = integer. Notice that for fixed j and j', there are in general a large number of multiplets $\{j, j'\}$, except for $\{M/2, 0\}$ and $\{0, M/2\}$, each of which occurs only once. For the former, the lowest corner is the point A where $N_a = N_b = 0$ which is a single state. For the latter, the lowest corner is B where $N_a = 0$, $N_b = M$ which is also a single state.

tions, so that the problem reduces to that of N_b noninteracting fermions. One can thus trivially write down the eigenstates of H' and $\mathbf{P}'$ in momentum space. There are $\binom{M}{N_b}$ such states. Operating with η^+ and ζ^+ on these states generates $\binom{M}{N_b}$ multiplets $\{j, j'\}$. Now obviously

$$j = \frac{1}{2}(M - N_b)\ , \quad j' = \frac{1}{2}N_b\ .$$

Thus we can easily write down explicitly the eigenfunctions for H' and $\mathbf{P}'$ for $\binom{M}{N_b}$ multiplets $\left\{\frac{1}{2}(M - N_b), \frac{1}{2}N_b\right\}$. The total number of such states is $\Sigma\binom{M}{N_b}(M - N_b + 1)(N_b + 1)$, where the summation extends from $N_b = 0$ to M. The summation is equal to $2^{M-2}(M^2 + 3M + 4)$. This is an enormous number of eigenstates, but still very small compared to the total number of eigenstates which is 4^M. We remark here that the eigenstates ψ_N of Ref. 1 are special cases of the states discussed in this section.

The eigenstates of H' constructed above obviously do not depend on W and are simultaneous eigenstates of T' and V'. We believe they are the only W-independent eigenstates of H', but we do not know how to prove this statement except in special cases.

(4) *ODLRO* — We shall show

Theorem 3. For any state ψ for which $j^2 - j_z^2 = O(M^2)$, there is ODLRO.

The 2-particle reduced density matrix ρ_2 has matrix element

$$\langle b_s a_s | \rho_2 | b_r a_r \rangle = \psi^+ a_r^+ b_r^+ a_s b_s \psi \ .$$

Thus

$$\Sigma e^{i\pi\cdot(\mathbf{r}-\mathbf{s})} \langle b_s a_s | \rho_2 | b_r a_r \rangle = \psi^+ \eta^+ \eta \psi = \psi^+ (J_x + iJ_y)(J_x - iJ_y)\psi$$
$$= j^2 - j_z^2 + j + j_z \ .$$

Using

$$\langle b_{r'} a_r | \phi \rangle = M^{-1/2} e^{i\pi\cdot\mathbf{r}} \delta(\mathbf{r} - \mathbf{r}')$$

as a trial wave function for ρ_2, we find the expectation value of ρ_2 to be

$$\langle \rho_2 \rangle = \frac{1}{M}(j^2 - j_z^2) + O(1) = O(M) \geqq 0.$$

Thus the largest eigenvalue of ρ_2 is $O(M)$ and the state has ODLRO.[2]

In Ref. 1 we had showed that the states ψ_N have ODLRO. That fact is a special case of the above theorem, because for ψ_N, $j = M/2$, and $j_z = -M/2 + N$.

In the above discussions, the pairs are particle-particle pairs. If the particle is charged e, then the state exhibits[2] flux quantization in units of $ch/2e$. If $j'^2 - j_z'^2 = O(M^2)$, the system exhibits particle-hole ODLRO. There is no superconductivity for such a system.[2,3] Thus *j is related to superconductivity and j' to magnetic properties.*

(5) *Unitary Operators U_b and X* — We define these two operators as follows:

$$U_b a_r U_b^{-1} = a_r \ , \quad U_b b_r U_b^{-1} = e^{i\pi\cdot\mathbf{r}} b_r^+ \ , \quad U_b^2 = 1 \ , \tag{11}$$

and

$$X a_r X^{-1} = e^{i\pi\cdot\mathbf{r}} a_r \ , \quad X b_r X^{-1} = e^{i\pi\cdot\mathbf{r}} b_r \ , \quad X^2 = 1 \ . \tag{12}$$

Operator X is well known and operator U_b has been discussed in the literature.[4] We observe that

$$U_b b_{\mathbf{k}} U_b^{-1} = b_{\pi-\mathbf{k}}^+ \ , \tag{13}$$

and

$$\zeta = U_b \eta U_b^{-1} . \tag{14}$$

Theorem 4. Writing $H'(W)$ for H', we have

$$U_b H'(W) U_b^{-1} = H'(-W) , \tag{15}$$

$$U_b (\Sigma b^+ b) U_b^{-1} = M - \Sigma b^+ b , \quad U_b (\Sigma a^+ a) U_b^{-1} = \Sigma a^+ a . \tag{16}$$

Theorem 5.

$$X H'(W) X^{-1} = -H'(-W) , \tag{17}$$

$$X(\Sigma a^+ a) X^{-1} = \Sigma a^+ a , \quad X(\Sigma b^+ b) X^{-1} = \Sigma b^+ b . \tag{18}$$

It follows that

$$\begin{aligned}
(XU_b)(H'(W))(XU_b)^{-1} &= -H'(W) , \\
(XU_b)(\Sigma a^+ a)(XU_b)^{-1} &= \Sigma a^+ a , \\
(XU_b)(\Sigma b^+ b)(XU_b)^{-1} &= M - \Sigma b^+ b .
\end{aligned} \tag{19}$$

Denoting by Spm (W, N_a, N_b) the spectrum of $H'(W)$ for given N_a and N_b, we have, by Theorem 4,

Theorem 6.

$$\begin{aligned}
\text{Spm}\,(W, N_a, N_b) &= \text{Spm}\,(-W, N_a, M - N_b) \\
&= \text{Spm}\,(-W, M - N_a, N_b) \\
&= \text{Spm}\,(W, M - N_a, M - N_b) .
\end{aligned} \tag{20}$$

By Theorem 5, we have

Theorem 7.

$$\text{Spm}\,(W, N_a, N_b) = -\text{Spm}(-W, N_a, N_b) . \tag{21}$$

Combining these two results we obtain

$$\begin{aligned}
\text{Spm}\,(W, N_a, N_b) &= -\text{Spm}\,(W, N_a, M - N_b) \\
&= -\text{Spm}\,(W, M - N_a, N_b) \\
&= \text{Spm}\,(W, M - N_a, M - N_b) .
\end{aligned} \tag{22}$$

(6) *Limit* $M \to \infty$ — We shall now put $\varepsilon = 1$ in (2). Diagonalizing $\mathbf{J}^2$, $\mathbf{J}'^2$, J_z, J'_z, H', $\mathbf{P}'$, we have also diagonalized N_a and N_b because of (10). Let the lowest eigenvalue of H' at a fixed N_a, N_b be denoted by $E_0(W, N_a, N_b)$. Now keeping fixed the values of

$$N_a/M = \rho_a \ , \quad N_b/M = \rho_b$$

we approach the limit $M \to \infty$. It can be proved, by a method used in Ref. 5, that $M^{-1}E_0$ approaches a limit which we shall denote by $f(W, \rho_a, \rho_b)$. f is the lowest eigenvalue of H' per site at fixed densities ρ_a and ρ_b.

The function f has many symmetries. Because of Theorems 1 and 2,

$$f(W, \rho_a, \rho_b) = f(W, \rho_b, \rho_a) = f(W, 1 - \rho_a, 1 - \rho_b) = f(W, 1 - \rho_b, 1 - \rho_a) \ . \tag{23}$$

Because of (20),

$$f(W, \rho_a, \rho_b) = f(-W, \rho_a, 1 - \rho_b) = f(-W, 1 - \rho_a, \rho_b) \ . \tag{24}$$

These symmetries are illustrated in Fig. 2.

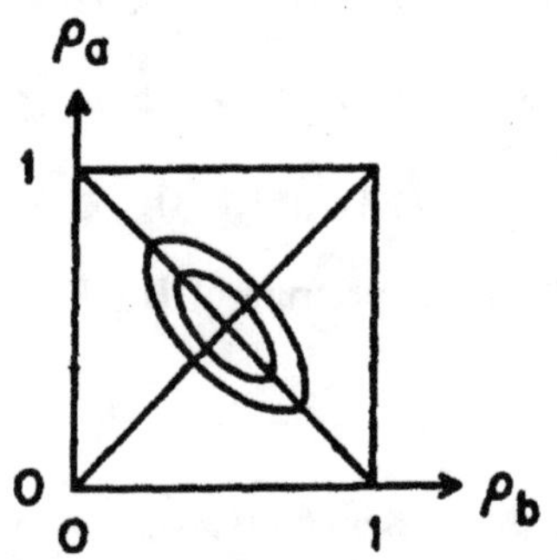

Fig. 2. Equi-f contours in ρ_a, ρ_b plane (schematic). Because of (23), these contours are reflection symmetrical with respect to the $\rho_a = \rho_b$ axis and the $\rho_a + \rho_b = 1$ axis. Because of Theorem 8, these contours are convex. One can obtain the $(-W)$ contours from the (W) contours by a rotation through 90° around the center of the square.

Theorem 8. $f(W, \rho_a, \rho_b)$ as a function of ρ_a and ρ_b is continuous and concaves upwards.

Theorem 9. $f(W, \rho_a, \rho_b)$ as a function of W concaves downwards.
These two theorems can be proved using the methods of Ref. 5.

Theorem 8 and Eq. (23) show that the minimum of $f(W, \rho_a, \rho_b)$ for fixed W is $f(W, 1/2, 1/2)$. This minimum value may be shared by f at other values of (ρ_a, ρ_b) than $(1/2, 1/2)$. Let the region of (ρ_a, ρ_b) where this is true be denoted by R, and call the states that have this minimum value of f *lowest states.* (23) shows that R is reflection symmetrical with respect to the axis: $\rho_a = \rho_b$, and with respect to the axis: $\rho_a + \rho_b = 1$. Using Theorem 8 we can show

Theorem 10. The region R in (ρ_a, ρ_b) where $f(W, \rho_a, \rho_b) = f(W, 1/2, 1/2)$ is convex. Possible schematic shapes of R are illustrated in Fig. 3.

Each of the *lowest state* belongs to a multiplet $\{j, j'\}$. Within that multiplet the leading state (i.e. where $j_z = j, j'_z = j$,) is also a *lowest state*. Hence it must be in the $j_z \geqq 0, j'_z \geqq 0$ quadrant of R. Thus

Theorem 11. All the *lowest states* on the boundary of R have $j = |j_z|, j' = |j'_z|$. Finally we remark that for the points $\rho_a = 0$ (or $\rho_b = 0$,) the system is devoid of a (or b) particles. Hence the value of $f(W, 0, \rho_b)$ and $f(W, \rho_a, 0)$ can be easily evaluated. (23) then allows one to write down $f(W, 1, \rho_b)$ and $f(W, \rho_a, 1)$. Thus the value of f on the boundary of the square in Fig. 2 is known.

We now define $g(W, \rho_a, \rho_b)$ to be highest eigenvalue of H' per site. Equation (22) then shows that

$$g(W, \rho_a, \rho_b) = -f(W, \rho_a, 1-\rho_b) = -f(W, 1-\rho_a, \rho_b) . \tag{25}$$

More generally we define the free energy per site by

$$F(\beta, W, \rho_a, \rho_b) = \lim(-M\beta)^{-1} \ln(\text{p.f.}) \tag{26}$$

where

$$(\text{p.f.}) = \text{trace of block of } \exp(-\beta H') \text{ belonging to given } \rho_a, \rho_b , \tag{27}$$

and the limit is for $M \to \infty$. Then

$$F(\infty, W, \rho_a, \rho_b) = f(W, \rho_a, \rho_b) ,$$
$$F(-\infty, W, \rho_a, \rho_b) = g(W, \rho_a, \rho_b) . \tag{28}$$

The function F has many symmetries. Theorems 1 and 2 show that

$$\begin{aligned} F(\beta, W, \rho_a, \rho_b) &= F(\beta, W, \rho_b, \rho_a) \\ &= F(\beta, W, 1-\rho_a, 1-\rho_b) \\ &= F(\beta, W, 1-\rho_b, 1-\rho_a) . \end{aligned} \tag{29}$$

Equation (20) shows that

$$F(\beta, W, \rho_a, \rho_b) = F(\beta, -W, \rho_a, 1-\rho_b) = F(\beta, -W, 1-\rho_a, \rho_b) . \tag{30}$$

Equation (21) shows that

$$F(\beta, W, \rho_a, \rho_b) = -F(-\beta, -W, \rho_a, \rho_b) . \tag{31}$$

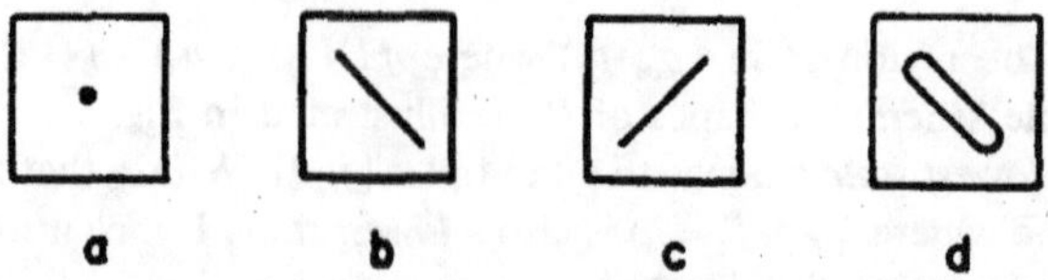

Fig. 3. Possible shapes for R. R is convex and is reflection symmetrical with respect to the $\rho_a = \rho_b$, and the $\rho_a + \rho_b = 1$ axes. For case c there is particle-particle ODLRO at low temperatures in the open line segment. For case d there is particle-particle ODLRO at low temperatures *inside* of the region R. These cases exhibit superconductivity.

These two last equations together show that

$$F\left(0, W, \rho_a, \frac{1}{2}\right) = F\left(0, W, \frac{1}{2}, \rho_b\right) = 0 \,. \tag{32}$$

Acknowledgments

One of us (CNY) is supported in part by the National Science Foundation under grant number PHY 8908495.

References

1. C. N. Yang, *Phys. Rev. Lett.* **63** (1989) 2144.
2. C. N. Yang, *Rev. Mod. Phys.* **34** (1962) 694.
3. W. Kohn and D. Sherrington, *Rev. Mod. Phys.* **42** (1970) 1.
4. H. Shiba, *Prog. Theor. Phys.* **48** (1972) 2171.
5. C. N. Yang and C. P. Yang, *Phys. Rev.* **147** (1966) 303.

www.ingramcontent.com/pod-product-compliance
Lightning Source LLC
LaVergne TN
LVHW010334220926
848484LV00004BB/896

9789819826629